Risk Assessment

— and —

Decision Making

IN BUSINESS AND INDUSTRY

A Practical Guide — Second Edition

Risk Assessment

and

Decision Making

IN BUSINESS AND INDUSTRY

A Practical Guide — Second Edition

GLENN KOLLER

Chapman & Hall/CRC
Taylor & Francis Group
Boca Raton   London   New York   Singapore

Published in 2005 by
Chapman & Hall/CRC
Taylor & Francis Group
6000 Broken Sound Parkway NW
Boca Raton, FL 33487-2742

International Standard Book Number-10: 1-58488-477-0 (Hardcover)
International Standard Book Number-13: 978-1-5848-8477-4 (Hardcover)
Library of Congress Card Number 2004061808

Library of Congress Cataloging-in-Publication Data

Koller, Glenn R. (Glenn Robert), 1951—
        Risk assessment and decision making in business and industry : a practical guide / Glenn Koller. — 2nd. ed.
        p. cm.
        Includes bibliographical references and index.
        ISBN 1-58488-477-0 (alk. paper)
        1. Rick assessment. 2. Decision making I. Title.

HD61.K63 2005
658.15'5—dc22   2004061808

Dedication

To Karen —
Without you none of this would be possible.
To Stephanie, Kat, and Matthew —
Your support is never taken for granted.

Contents

About the Author

Dr. Glenn Koller is senior advisor for risk evaluation at a major energy company. He has responsibility for aspects of risk and uncertainty analysis, management, and training. He received his Ph.D. in geochemistry/geophysics from Syracuse University. Responsibilities of his current position include implementation of risk-assessment and risk-management technologies and processes in the corporation; development of statistical routines that comprise corporate risk systems; keeping abreast of risk technology developed by vendors, other companies, and the national laboratories; marketing risk technology; and performing technical and consulting services. Areas of responsibility include portfolio management, business and product development, environmental concerns, ranking and prioritization of projects and products, analysis of legal scenarios, and other aspects of the corporation's diversified business. Glenn has authored several books on risk/uncertainty (see www.risksupport.com) and currently resides in Tulsa, Oklahoma. Glenn's email address is riskaid@cox.net.

Preface

This book, like the original edition, is a hybrid — a confluence of technical aspects and advice regarding how risk/uncertainty (R/U) techniques and processes can be practically implemented in the corporate world. All information is described in layman's terms so anyone can pick up this treatise and comprehend the content.

As was the original edition, this book is comprehensive, covering four broad aspects of R/U, those being (1) general concepts, (2) implementation processes, (3) technical aspects, and (4) examples of application. Practical guidance is given in each area and the book leads the reader logically through the R/U landscape beginning with why R/U is useful and when, defining the basic components and processes, describing in layman's terms the technical aspects, describing in detail corporate (organizational, cultural, etc.) and human considerations and impediments to R/U process implementation, and culminating in examples of real business applications.

A main area of focus in the first edition was the general processes involved in quantitatively assessing risks and their impact on value. This new edition builds on those principles, culminating in a description of two critical probabilistic measures of project value — the expected value of success (EVS) and the expected value for the portfolio (EVP). These two probabilistic measures of value can be significantly different in magnitude. The EVS finds utility in project assessment and management. It is the value upon which project construction, logistics, contractual, marketing, and other decisions are based. The EVP is critical for making decisions at the portfolio level. It is the value that should be used in the summation of individual project values to arrive at portfolio value and for ranking projects within a portfolio. Both metrics take advantage of a holistic assessment of risk and uncertainty as well as incorporation of chances of abject failure associated with any potential investment or portfolio of investments. How these metrics are employed in individual project and in portfolio management is clearly delineated and demonstrated.

While the aforementioned aspects of this new edition are important, the most significant offering is the advice and guidance related to implementation of R/U processes in modern corporations. Definitions for the terms risk and uncertainty are for the first time offered in such a way that they can be adopted by all aspects of science and business. This is a crucial breakthrough.

The first edition mainly addressed the various positive characteristics of risk assessment and management. In this new work are conveyed the difficult aspects of, and impediments to, implementation of practical and cogent R/U processes in any organization. From the author's real-life experiences, practical advice is offered that address the many technical, cultural, organizational, political, and other barriers to process change and implementation. Issues such as resistance to change, turf battles, experts who are not, establishing credibility, the reward system, where and how to penetrate the organization, the "not invented here" syndrome, and many other characteristics of modern-day corporate life are addressed, and practical guidance is offered for confronting these potential hurdles.

In addition, in this new edition the author projects a view of R/U from the corporate decision maker's position, and addresses how to employ R/U techniques to set budgets, manage portfolios, value investments, and perform other activities fundamental to corporate life. Also considered are the practical problems faced by decision makers, such as determining how new information impacts the decision-making process, what questions to ask to determine the quality of information conveyed, and how new information and techniques can be integrated into the existing decision-making scheme.

The first edition of this book lays the technical and organizational groundwork necessary to implement R/U-based processes in a corporation. This new volume retains that information and expands on that foundation to address the many technical, organizational, cultural, political, and human-nature aspects of R/U process implementation. Guidance regarding how to calculate critical value-measuring metrics such as the EVS and EVP is conveyed. The integration of advice addressing technical and non-technical characteristics of R/U-related corporate life makes this book a complete primer for people in any aspect of modern-day business.

Acknowledgments

I thank Graham Cattell, Director of Major Projects, Group Technology for his review of this book and for granting permission to publish. Graham's direction and business focus are also greatly appreciated.

I acknowledge and thank Michael A. Long for his friendship and contributions to the development of our overall understanding of risk/uncertainty assessment and management, and for his prowess in software development. Mike's contributions to our effort cannot be understated.

Appreciation and thanks are extended to Steve Jewell for his camaraderie and his guidance and understanding in our journey to implement risk-related processes in the corporation. Steve's good humor and technical abilities are essential elements of our efforts.

Last, but certainly not least, I extend my heartfelt love and appreciation to my wife Karen and our children Stephanie, Kat, and Matthew. Without their understanding and support, monumental efforts such as the writing of these books would not be possible.

Introduction

It is not primarily about technology — it is about changing behaviors.

WHAT COMPANIES REALLY WANT TO DO

It is about money. Companies want to make it. Practically? Yes. Ethically? Certainly. As fast as possible? Of course.

When I wrote the first edition of this book (*Risk Assessment and Decision Making in Business and Industry: A Practical Guide*), I did what I thought was a credible job of describing the theories and processes practically to implement risk and uncertainty in the analysis of all sorts of problems. I even included a few relatively simple and easy-to-understand examples of such applications. As it turns out, this was equivalent to imparting to the new driver how the brakes work, how the engine works, the rules of the road, etc. However, having intimate knowledge of each individual aspect does not necessarily afford the practical ability to safely operate a car. Safe operation requires the integration of all individual facets.

It should be noted that later in this book there is an entire chapter devoted to this subject (and with the same title as this section). This is by design. It is important here to briefly denote what companies really want to do. Later, after the reader has supposedly plowed through the chapters between this place and the aforementioned chapter, I will use the information in the intervening sections of this book to illustrate by example what companies really want to do.

Following publication of the first edition of this volume, I received many inquiries regarding just how such techniques and processes would be implemented in an actual corporation and how these things can be used to reduce costs and/or increase revenue. This new volume, however, is *not* aimed at relating to the reader how to analyze an investment, for example, from the risk/uncertainty viewpoint. Excellent books such as John Campbell's *Analyzing and Managing Risky Investments* (2001) address such issues. Nor does this new volume address the issue of how to build risk models for myriad situations — my second book, *Risk Modeling for Determining Value and Decision Making* (Chapman & Hall/CRC Press, 2000), addresses that problem. Rather, the point of this new book is to focus on the practical integration of the many fundamental aspects of risk assessment and to impart to the reader how such an integrated approach can be successfully implemented in a small- or large-company environment for the purpose of increasing revenue, reducing cost, capturing more market share, and other typical business aspirations.

As previously stated, it is about money. It is also all about choices. What any company would like to do is to make smart decisions — to make smart choices. Smart typically is translated to mean something that is done to extend the life of the company. In turn, this implies that something has transpired to enhance company revenues.

Every revenue-enhancement and cost-reduction step involves a choice. Corporate executives sometimes are simply choosing between taking a step or not. For example, we might be trying to decide whether or not we will implement a work-from-home and job-sharing policy in an attempt to increase productivity. We either do it or we do not. Other choices involve one or more alternative investments. For example, we might be trying to decide in which of several cities we will open our new and quite expensive distribution center.

I have said that it is all about money and choices, but it also is all about something else: accounting for risk and uncertainty and practically establishing the risk/uncertainty process within the existing corporate organization and culture. Too often, a typical risk assessment/management process involves the listing of risks, a discussion of uncertainties, and the formulation of mitigation plans for the predominant risks. These are all fine and reasonable steps to take; however, the real benefit of a comprehensive and holistic risk assessment is mainly derived from impacting the value of any project — be it a marketing opportunity, an acquisition, a construction project, or whatever — with the risks and uncertainties. Just *how* we do this will be the subject of the bulk of this book. Just *why* we do this is to be able to make those smart decisions alluded to earlier.

Consider the situation in which corporate executives are struggling with a decision as to which of two countries will be the location of a major new and expensive facility. Marketing people have considered all of the demographic aspects of both countries. Similarly, logistics experts, commercial folks, construction engineers, and finance personnel all have assessed the situation in both arenas. The first problem that can typically arise is that most of these assessments arrive as separate reports. That is, a folder comes from the marketing people filled with text, plots, and data tables that relate their view of the two marketing scenarios. A file received from the logistics area imparts the problems and benefits related to the transportation networks and other logistical aspects of each country. And so it goes for all other facets of the analysis.

This is all great stuff, but usually it is not integrated. It is too much to ask that all of these separate entities get together and somehow build an integrated story. How would they do that? A related and equally untenable problem is that no matter the measure of value of the project — net present value, (NPV) internal rate of return, (IRR) discounted return on investment, (DROI) return on capital employed, (ROCE) or whatever — it has not been impacted (i.e., damped or enhanced) by the discipline-specific risks and uncertainties. To make a smart decision, corporate executives would best be served by comparing project values that reflected the impact of all of these considerations.

So, one focus of this book will be to demonstrate, in a modern corporate environment, just how risks and uncertainties can be used to impact the value of a project — whatever that project is. However, there is a second and equally salient risk/uncertainty consideration — that of ensuring that the assessment process is comprehensive.

Assume that corporate executives have followed the processes and techniques outlined in this book and have successfully impacted the cash flows and, let's say, the NPV of the project utilizing the risk/uncertainty considerations from marketing, logistics, engineering, finance, and commercial. They now have two risk-weighted NPV values to compare. However, in one country, the political risk is very great (unstable government and the possibility of an uprising). If political risk and uncertainty of this sort have not been properly integrated into the assessment and, in turn, have not been considered in the calculation of the risk-weighted NPV for investment in that country, the comparison of the two country NPV values can lead to an erroneous investment decision.

So, smart decisions and choices are facilitated by the following:

- Assuring that the risk/uncertainty assessment process is holistic and includes such considerations from *all* pertinent aspects
- Assuring that the value of a project, however calculated, has been properly impacted by all risks and uncertainties

Just how these things can be practically accomplished in a modern corporation will be the focus of the new part of this book. An entire chapter and other sections of this book are devoted to imparting to the reader the practical problems and difficulties — and some recommended solutions — related to implementing a risk/uncertainty process in an existing corporate organization and culture. The more technical aspects — the focus of the earlier volume — are also preserved and related.

PRACTICAL RISK ASSESSMENT

In the title of this book is the word practical. That word will serve as the guiding principle for this entire volume. It seems the bulk of books with statistical or risk assessment themes are written for other people in the field, or to impress those same people. This book is intended to impress the reader with these facts:

- You do not need to be a statistical guru or risk assessment expert to use risk assessment in your decision-making process.
- Risk assessment can be practically applied at any decision-making level and in any business, educational, or technical arena.
- Risk assessment is a process — not just a set of equations — the success of which mainly hinges on skillful handling of organizational, educational, communication, and political aspects of the process. Technical considerations, while important, are secondary.
- Risk assessment can be explained and implemented using plain, understandable terms and concepts. Nebulous and (almost always) self-serving technical jargon is not necessary and is, in fact, detrimental.

In more than 20 years of working in this field, I have come to find that success, however defined, stems mainly from bringing to bear finely honed interpersonal, communication, and educational skills. Technical competence, while required, is an empty promise when not combined with the knack for engaging and facilitating individuals and groups.

It also has been my experience that many risk assessment processes touted by consultants tend to be a day's worth of process packed into a week. I find that people at all levels, in whatever business, are busy — busier perhaps than any person in past times with an equivalent position. Taking more time than is absolutely necessary to design a risk process and to implement the resulting risk model is a destruction of value. Hopefully, the practical approach put forth in this book will, at least in part, preclude such practices.

Like marriages in the U.S., a great proportion of risk assessment projects undertaken either fail outright or are unable to reach their intended goal or full potential. Risk assessment efforts tend to fail for many of the same reasons as failed marriages. Bad assumptions, poor communication, lack of sufficient partner/stakeholder intimacy, selfishness and arrogance, and unrealistic expectations are just a few of the foibles that can individually or in combination torpedo the most astute and technically brilliant risk assessment efforts. In this book I will address the technical aspects of risk assessments of all types as well as consider in depth the organizational, political, and human issues that are most difficult to grapple with.

Risk assessment is but one part of the overall assortment of processes that comprise the art and science of decision analysis. It is beyond the scope of this book to address any but the risk-assessment-related aspects of decision making. Explanation of decision-making tools such as strategy tables and the like can be found in articles and books listed in the reference section for this introduction.

It is not the intent of this book to make risk assessment experts of those who now are not. Rather, the aim is to convey to the reader in plain English the fundamental principles of risk assessment and the enveloping processes. Although risk assessment software and experts still may need to be engaged to successfully implement a risk process in an organization, students of this book will understand risk-related processes and technologies. They will be able to discuss intelligently with those experts the unique considerations of their predicament and will be capable of interacting with the experts in the effort to solve the problem.

This book intentionally lacks a disciplinary slant. That is, in these pages you will find examples from these areas:

- Brokerage-house portfolio management
- Legal decision making
- Construction
- Oil/gas exploration
- Environmental assessments
- Engineering
- Marketing
- Government
- Manufacturing

Many other areas are represented as well. Risk assessment is a universally applicable technology and, as the following section will attest, it does not have to be hard to use.

YOU CAN DO IT

As mentioned previously, it is my job every day to bring the risk process and risk models to bear on all types of businesses, processes, and situations. If I took the traditional approach to this problem — that of requiring of users things such as esoteric statistical information that they are not prepared to supply — I would fail miserably.

The key to successful risk-process and risk-model implementation is to build a risk system that requires from the user almost nothing more than they can presently provide. If you expect that through zeal, fervor, osmosis, or a concentrated training effort you will convert all potential customers into rudimentary statistical beings, you are about to experience one of life's great disappointments.

When I think of a risk assessment, I do not have in mind only, for example, the engineer who is comfortable with spreadsheets, distributions, equations, and computer technology. Such individuals rarely can see how advanced they are in their thinking, abilities, and approach. They generally cannot appreciate why everyone else does not do it like they do it. (What is the matter with these other people? This is easy.) To me, implementing a risk process means designing a system that requires from the user very little, if anything, that he is not presently prepared to deliver. This can be a challenge, but it is a challenge that I have successfully met time and again. You can too.

Most of this book is dedicated to explaining just how you might go about achieving this goal. Spending time with the potential users of the system, understanding their business, facilitating the risk-model-building process and many other steps and processes are required to build a successful risk model. But the point is that it can be done.

The result of a properly orchestrated risk-model-building process is a value-added risk system that is practical and comprehensive. As emphasized by the title of this book, practical is the operative term. If the system resulting from the risk-process-building effort is not practical and easy for the non-risk-expert to use, then the effort is a failure. Execution of the steps outlined in this book should lead to a successful implementation philosophy and process. However, as the reader will see, it is not all "sweetness and light."

SELECTED READINGS

Campbell, J. M., *Analyzing and Managing Risky Investments, 1st Edition*, John M. Campbell, Norman, OK, 2001.

Koller, G. R., *Risk Modeling for Determining Value and Decision Making*, Chapman & Hall/CRC Press, Boca Raton, FL, 2000.

Schuyler, J. R., *Decision Analysis Collection*, John R. Schuyler, Aurora, CO, 2004.

Schuyler, J. R., *Risk and Decision Analysis in Projects*, Project Management Institute, Newton Square, PA, 2001.

Welch, David A., *Decisions, Decisions: The Art of Effective Decision Making*, Prometheus Books, Amherst, NY, 2001.

1 About Risk and Uncertainty

VARIOUS VIEWS ON THE SUBJECT — EINSTEIN WAS NEVER GOING TO GET THERE

As students of physics and others know, Albert Einstein was in pursuit of the Unified Field Theory. In a nutshell, he sought the unifying fundamental truth that underlay and explained all universal phenomena, such as gravity and light. I have studied the workings of corporations for a long time. If the mechanics of such organizations are *any* indication of how the universe operates, it is my opinion that Dr. Einstein was never going to get there. Of course, I jest about the link between organizations and the physics of the universe, but the risk/uncertainty (R/U) environment within most corporations can be relatively chaotic and is no joke.

Although outsiders typically construe corporations to be single unified entities, these organizations in effect are collections of potentially competing and relatively independent components. For example, a modern corporation might be comprised of departments such as the following:

- Finance
- Logistics
- Law
- Commercial
- Health and safety
- Human resources
- Construction
- Research and development
- Marketing
- Retail
- Advertising
- Public relations
- Security
- Planning
- Environmental

And there are many more general and specialty departments, depending on just what the corporation actually does. With regard to R/U, each of these corporate components can harbor views and practices on R/U that are not only different, but can be diametrically opposed.

Personnel in the health and safety department, for example, might view risk as a bad thing. That is, anything perceived to be a risk to the health and safety of company employees or to the public is to be eradicated or the frequency of occurrence and impact of such a risk are to be reduced as much as possible. Uncertainty is perceived to be something to be diminished or eliminated.

Representatives from finance can have a different view. Making risk/reward evaluations of investments and projects is part and parcel of what they do. Greater risk generally (hopefully) yields greater returns. The folks in finance do not typically attempt to take on more risk than they think the corporation can handle (i.e., diversify, hedge), but their mission is to select opportunities that represent the best chance for relatively high returns on investment. Such opportunities include those that have relatively high levels of risk. In addition, volatility (i.e., uncertainty) can be seen as an opportunity for profit and, therefore, might be something to be pursued and embraced.

So, the views of R/U in a corporation can be all over the map, as can be the methods used by various groups to express their perception of R/U. Some examples follow.

One department might use text to convey its assessment of R/U. For example, "It is our opinion that the most critical risks are those that are political and those that are associated with logistics. The political challenges stem from the attitude of the host government regarding...."

Another department might employ the risk matrix methodology as shown in Figure 1.1. Another department might use a risk-register approach as illustrated in Figure 1.2. Simulation might be the choice of another department. Output from a simulation is shown in Figure 1.3. Yet another department might choose to use the colored-traffic-light approach to express its view on risk, as depicted in Figure 1.4.

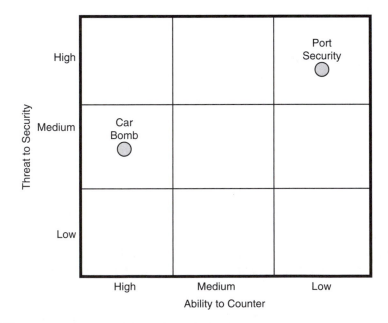

FIGURE 1.1 Example of the risk matrix method.

RISK	DEFINITION	MITIGATE ACTION	PERSON ACCOUNTABLE	DATE	MIN. $	M.L. $	MAX. $	PRIORITY
Cont. Avail.	Click here for full expl.	Early negotiation	Joe B.	02/05	$0.7K	$1.2K	$1.5K	H
Permits	Click here for full expl.	Process — early start	Sally S.	05/05	$1.2M	$1.5M	$1.7M	M

FIGURE 1.2 Example of a risk register.

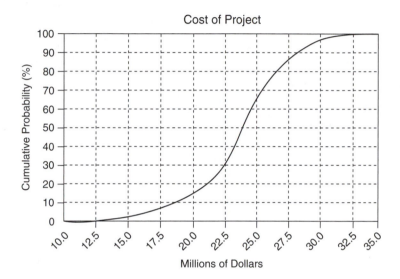

FIGURE 1.3 Cumulative frequency curve resulting from a Monte Carlo simulation.

Risk/reward plots, colored maps, and many other approaches might be employed. It is the task of a skilled risk-modeling facilitator to integrate these various expressions and representations into a common metric that can be used to impact the perceived value of an opportunity or liability. Techniques to facilitate this will be delineated in later chapters of this book. Individual techniques (risk matrix, simulation, risk registers, and so on) are also addressed in detail in later chapters. The point is, do not expect consistency or uniformity. With respect to views and expressions of R/U in any corporation, chaos is the rule rather than the exception.

DEFINITIONS OF RISK AND UNCERTAINTY

Building on the chaos theme, one should neither be surprised nor discouraged by the broad perception of R/U in a corporation and the myriad definitions for the terms. As illustrated in the previous section, consensus is nearly nil regarding even the fundamental concept of whether or not risk represents something to be eradicated or embraced. Just as an example, in finance alone, risk can be defined in the following ways, among others:

Risk	Definition	Ability to Mitigate
Acreage	Bid for acreage unsuccessful	⬤ Green
Partner	Critical partner drops out	◯ Yellow
Environmental	Environmental problems cancel project	⬤ Red

FIGURE 1.4 Example of traffic light format.

- Financial risk: the chance of exposure to monetary loss
- Specific risk: the risk associated with a particular company or a specific financial instrument
- Market risk: risk associated with the field in which a company acts or in which a financial instrument is traded

Having had intimate encounters in various industries with all of the departments (and more) listed in the previous section, I came to the conclusion early on that what was sorely needed was a definition of risk and uncertainty that could be at least understood — if not adopted — by the disparate entities that comprise a corporation. The definitions delineated below, I have found, can be applied across the broad spectrum of organizations.

Although the title of this section indicates that I will discuss the definitions of the terms risk and uncertainty, I will take the liberty to extend consideration to four terms:

- Risk
- Uncertainty
- Probability (sometimes termed chance of or likelihood)
- Consequence (sometimes termed impact)

There seems to be a propensity to confuse probability and/or uncertainty with risk. Risk, probability, uncertainty, and consequence are four distinct entities. For example, some folks like to consider risk to be the chance that something would happen (probability). Yet, when asked, "What are the risks to this project?" they do not reply, "25, 16.5, and 32." They always respond, "Well, the risks are that we will not have enough raw materials available to…." That is, they respond with definitions of the problems, not a list of percent-based probabilities.

So, I define risk to be:

A pertinent event for which there is a textual description.

The textual description certainly can contain numbers: "The risk is that the pipeline will experience pressures greater than 2500 psi for a period exceeding 2 days…" Under this definition, a risk could be an event that has a positive impact.

For example, in the finance world, people might identify a risk as: "There is a risk that taxes will be lower than expected…" or ""The risk exists that a favorable legislation will be passed…" Now, some purists do not like to consider that risks can be either negative or positive. The definition proposed here allows each risk-defining entity to decide for itself whether or not risks should be negative-impacts only, or, can include positive aspects.

Consider a typical construction project. Many things can, for example, threaten the schedule for the project. Severe penalties for late completion might be written into the contract, so the construction company is mightily concerned with items that could potentially retard execution. The company might make a list of perceived threats such as:

- More-than-usual amount of bad weather
- Union strike
- Permit delays
- Contractor availability (lack therof)

and so on.

In this case, a risk is that there will be more bad weather than normal. A risk is that there will be a strike by a union. A risk is that the issuance of necessary permits will be unacceptably delayed. A risk is that necessary contract labor will not be available in a timely manner. So, the risks are simply events for which there are textual definitions/descriptions.

Associated with each risk are typically at least two parameters:

- Probability of occurrence
- Consequence (impact) of occurrence

About each of these attributes, we could be sure. For example, if we flip a coin, we are sure that the probability of "heads" is 50%. We might also be sure about the prize of winning (i.e., if we get "heads" then we get to keep the coin). So, in this case, there is no uncertainty about either the probability or the consequence. However, in most business situations, we cannot state with 100% certainty just what the probability will be nor what the consequence will be if the risk is realized. This brings us to uncertainty.

Uncertainty, by my definition, can at least be related to probability and impact of a risk — and other parameters if appropriate. Uncertainty is typically, but not always, represented as a range of values (sometimes as a distribution) that encompasses the range of possibilities. So, uncertainties related to the construction example above might be as follows:

- There is uncertainty regarding the probability of an abnormal amount of foul weather (range of percentage chance we will experience more bad weather than normal)
- There is uncertainty regarding the duration of the abnormally bad weather (range of days in excess of normal number of bad-weather days)

- There is uncertainty regarding the cost per day of bad weather (range in dollars per day)
- There is uncertainty regarding the probability of a union strike (range of percent chance that we will experience a union strike)
- There is uncertainty regarding the duration of the strike (range of number of days)
- There is uncertainty regarding the cost per day of the strike (range in dollars per day)
- There is uncertainty regarding the probability that we will experience permit delays (range of percentage chance delays will occur)
- There is uncertainty regarding the duration of the permit delays (range in number of days)
- There is uncertainty regarding the cost per day of permit delays (range in dollars per day)
- There is uncertainty regarding the unavailability of required contract labor (range of percent chance we will not get timely access to required labor)
- There is uncertainty regarding the delay caused by labor unavailability (range in days of delay)
- There is uncertainty regarding the cost per day of labor delay (range in dollars per day)

So, we can simply calculate the impact of, for example, the bad-weather risk by multiplying:

- The percentage chance of an abnormally long stretch of bad weather times
- The number of days times
- The cost per day

If uncertainty is involved, this multiplication might be performed by a Monte Carlo process that allows combinations of distributions. Similar calculations can be executed for the other risks. When we have identified our risks and uncertainties, we will take steps to mitigate the risks. For example, we might consider building shelters for some projects so they can carry on in inclement weather. This reduces the probability of a delay due to bad weather (but not the consequence of such a delay — that is, if we are in fact delayed, the cost per day of delay has not been reduced). So, we do strive to reduce probability and consequence.

When probability and/or consequence related to a risk are reduced to the point that we no longer consider them to be a threat to success, the risk is eliminated — that is, it is taken off the list. So, we do *not* strive to reduce risks — only to eliminate them from the list. We *do* strive to reduce the probability and the consequence, hopefully to the point that the risk can be eliminated.

Uncertainty can also be reduced, but it is not necessarily associated with a reduction (or increase) in probability or consequence. For example, one risk to an engineering project might be the gap between two physical elements. If the gap is about one foot, then we have excellent physical means of measuring that gap and the uncertainty resulting from multiple measurements of the same gap is that of measurement error associated with the process. Relative to the gap, this error represents a very small range of uncertainty. However, if the gap is exceedingly small and requires a relatively indirect means of measurement (x-rays, electron microscopes, etc.), then, relative to gap size, the range of uncertainty can be large.

Consider a situation on an offshore oil rig. Directly below the rig that floats on the ocean surface are exceedingly expensive sea-floor installations. This sea-floor equipment is vulnerable to damage from items accidentally dropped from the rig. Given the current rig design and rig-floor procedures and processes, it has been estimated that the probability of such an accidental drop — based on historical data from other such installations — is in the range of 35 to 50% within a year's timeframe. This is a 15% range of uncertainty.

New procedures aimed at reducing the probability of accidental drops are about to be put in place. These are new processes and procedures that have not been previously tried. Management estimates that the new processes and procedures will reduce the probability of a drop to about 5 to 25% over the course of one year. This is a significant reduction in the probability of a drop, but the uncertainty range — because these new processes and procedures are untried — is now 20%. So, although the mean probability of a drop has been reduced, the uncertainty related to a drop has increased.

So, once again, there are four distinct concepts and terms here — risk, uncertainty, probability, and consequence. A risk typically has an associated probability of occurrence and at least one associated consequence. About these probabilities and consequences we can be certain or uncertain. Uncertainty is typically a range of values, sometimes represented as a distribution. The range of uncertainty is not necessarily directly related to an increase or decrease in probability or consequence. We strive to reduce the probability, consequence, and uncertainty. When either or both probability or consequence have been reduced to levels that we deem to be acceptable, we can entertain removing the risk (a pertinent event for which there is a textual description) from our list of considerations.

Decision processes are impacted by application of both the risk and uncertainty concepts. In facilitated workshops, the teasing out of all pertinent major risks can give decision makers a more realistic view of a project. Even if no numbers are assigned to the risks (i.e., no associated probabilities or impacts), the basic task of delineating all applicable major risks and creating a written list promotes appropriate conversations. Such lists of risks highlight the importance of engaging personnel from disciplines whose activities are not typically considered to be correlated. For example, such a list might point out that the legal and commercial staff might need to confer. In addition, the creation of a physical written list can convey to decision makers the breadth and magnitude of projected threat from risks. When no list is kept, it is difficult to compile mentally the number and magnitude of the risks, much less to recognize correlations and synergies.

As is demonstrated in later chapters of this book that deal with the concepts of distributions and uncertainty, recognition of the range of possible inputs or outcomes from a project can dramatically impact the decision-making process. In many situations, even if probabilistic analyses have been performed, decision makers often are more comfortable having delivered to them a deterministic (i.e., single-valued) result. This can be a P50 value or a mean or whatever they choose. However, two greatly different projects can reflect the same deterministic result.

For example, project A might have a resultant profit range from $40M to $60M with a mean profit of $50M. Project B might show a profit that ranges from $20M to $75M but, because of the profit distribution shape, also has a mean profit value of $50M. Considering simply the two deterministic means is, in this simple case, not telling the complete story. Consideration of the range of uncertainty impacts the decision-making process. It has been my experience that presentation of standard deviation values and multiple P values (P20, P50, P80, etc.) are exceedingly poor substitutes for the presentation of the full range of possibilities — usually presented as a frequency or cumulative frequency curve (these types of data displays are considered in detail in later chapters of this book).

Consideration of uncertainty has other beneficial effects. Often, critical considerations in a project go unspoken or are taken for granted. When each element of the project is specifically considered as uncertain and appropriate conversations are had, a better view of the project and more appropriate decisions are likely.

TWO VALUES — THE EXPECTED VALUE OF SUCCESS (EVS) AND THE EXPECTED VALUE FOR THE PORTFOLIO (EVP)

When constructing the outline for this book, I struggled much with just where to insert this section. Because the expected value of success (EVS) and expected value for the portfolio (EVP) precepts incorporate some technologies and concepts that have yet to be introduced in this book, I lamented over early introduction of EVS and EVP. However, these concepts are integral components of many of the methods and corporate traits described in the following text. Therefore, I decided that at least a rudimentary introduction to the EVS/EVP concept was warranted at this early stage. In following sections and chapters of this book, the EVS/EVP concept will be fleshed out using real-life business situations. The overview here is admittedly simple but does, I think, serve a purpose.

The fundamental concept behind the EVS and EVP is that each opportunity to be technically or financially evaluated might be represented by two measures of its value. To convey the basic principles, I typically employ the elemental example of a portfolio of 10 dimes (Figure 1.5).

Consider that each dime represents a project that might be executed. In this case, execution of the project is the flipping of the dime. Consider also, for the sake of simple arithmetic, that the probability of executing the project (or executing the project such that its impact will be felt in the timeframe currently being considered) is 50%. This is distinct from the 50% probability associated with the flipping of the coin. Note that the "S" in EVS indicates that we have been successful in executing the project and does *not* mean that we have executed a successful project — there is a difference. This difference will be demonstrated in the EVS/EVP example in Chapter 17. Let us further consider that if we do execute the project (flip the dime), heads will generate a return a bit greater than 10 cents (say, 11 cents) and tails will yield a return a bit less than 10 cents (say, 9 cents).

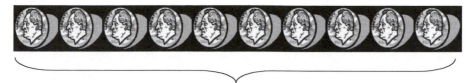

Portfolio of 10 independent projects

FIGURE 1.5 Portfolio of dimes — each dime representing a project to be executed.

So, given our 9- and 11-cent flip-of-the-dime results and the associated 50% probabilities, the EVS is 10 cents (the sum of 0.5 times each return). Independent of this, there is a 50% probability of the project not being executed or being executed outside our considered timeframe (our chance of abject failure). If this happens, the project yields nothing. The EVP value is calculated simply by adding the product of the probability of success (0.5) and the EVS (10 cents) to the product of the probability of abject failure (0.5) and the consequence of failure (0). The EVP in this simple case is 5.

If the engineer for a given project asked the project manager just how big a vessel he would have to build to hold potential winnings, the answer has to be a vessel with enough capacity to accommodate 10 pennies. So, the expected value of success (EVS) — the size of the prize if we are successful — is 10. However, when the engineer is replaced by a corporate representative in the project manager's office and the representative asks just how much this particular project would contribute to the corporate-portfolio project-output sum, the response has to be 5. So, the sum of all EVPs from all 10 — in this case, identical — projects would yield a portfolio value of 50.

Now, this is not rocket science. However, this example does introduce the concept that each project can be represented by two separate and equally important values, the EVS and the EVP. The EVS is important because, in a real-world project, we need to know what the project would yield if successful. Engineers need to know how big to design and build; personnel concerned with logistics need to know what type of contracts they need to sign to move the volume of product expected from a successful project; and so on. As will be demonstrated in later examples, the EVS can be, for example, the mean of a range of values that represent the combined uncertainty of the EVS components.

The EVP is the EVS damped by the combined chances of abject failure. As in the dime example, the EVS is 10. The EVS impacted by the chance of abject failure (10 x 0.5) yields an EVP of 5. This is the amount that the project (i.e., dime flip) would contribute to a portfolio of projects on a fully risk-weighted basis. Projections of portfolio yield or cost are summations of the EVP values for all of the projects that comprise the portfolio. As will be described in detail in later sections of this book, the EVS/EVP concepts, though simple, can be surprisingly difficult to implement in the modern corporation.

RISK ASSESSMENT AND RISK MANAGEMENT

Many R/U consulting firms consolidate their risk-related efforts under the general banner of risk management. I am not a believer in this generalization and draw a distinction between risk assessment and risk management. In this section I illustrate this distinction and why it is important in the corporate environment.

When I teach classes, I illustrate the unique aspects of assessment and management using a simple and, yes, crude example. I ask the class to envision a yard that I have to traverse. In this yard lives a vicious dog. What do I do? First, I assess the situation by seeking answers to certain probing questions:

- How fast is the dog?
- How fast am I?
- When does the dog sleep?
- Is he a light sleeper?
- How far is it across the yard?

When I have answers to these questions, I have assessed the situation. Now, what do I do about it? This is the management part. I might do any of the following:

- Try to cross the yard under the cover of darkness
- Try to outrun the dog
- Cross the yard when the dog is sleeping
- Buy some stilts and walk across the yard
- Drug the dog (no letters, please)

Although this is a silly little conundrum, it does illustrate the difference between assessing and managing the situation. People in my business are primarily involved with the assessment phase. As you will see in the following chapters, I advocate first trying to define just what is the question, then defining variables, then gathering data on uncertainty and chances of abject failure, then integrating the data. This is all about assessment. When the problem has been thoroughly analyzed and resultant plots and data generated, I typically consider myself to have reached the terminus of my expertise.

As illustrated above using the dog-in-the-yard example, management, by the definition promoted here, is mainly concerned with what to do about a situation after it has been exhaustively assessed. The steps of risk response and risk control are fundamental management steps. A person's or a company's aversion to risk (utility theory), available physical resources, cash reserves and credit line, political realities, partnerships, reputation concerns, and countless other tangible and intangible considerations influence the steps considered in the management process. It is beyond the ken of this author and, therefore, of this book to address the nuances of the risk management process. The reader is directed to some excellent texts on this subject such as those listed in the selected readings section at the end of this chapter. The focus of this book is on the process and techniques of risk assessment as it is defined here.

WHAT QUESTIONS DO I ASK AND HOW DO I INTEGRATE THE INFORMATION?

Being a decision maker is a precarious position. Just one potentially dire aspect of sitting in the decision-making seat is the assumption of responsibility for the outcome of a given decision. Another vexing aspect of occupying the decision-making seat is that individuals in this situation are assailed from all directions with data of varied nature and quality — not to mention the sometimes dubious characters delivering such information.

When receiving information upon which to base a decision, the decision maker has certain requirements related to the information and to the people ("them" in the discussion below) delivering the information. Among these are the following.

The decision maker needs them to deliver the message in jargon he can synthesize and can pass along. It is typical to receive data as engineering or science speak. Some decision makers might themselves be engineers or scientists, so they are not intimidated by such jargon. However, the decision maker typically has to pass along this information to people at higher levels or, at least, document the basis for the decision in a brief. This requires translation of the information into layman's terms. Such translation is best done prior to delivery of information to the decision maker so that conversations about the information can ensue. This is tough enough in the deterministic world, but is at least doubly daunting when grappling with the additional layer of obfuscation brought on by probabilistically-based (i.e., risk-based) data.

Decision makers need them to deliver a succinct answer or response. Individuals and groups delivering the information are usually justifiably proud of their accomplishments. As a result, it is not uncommon for them to include many ancillary details and bits of information that simply create an erudite haze that obscures the salient points. A succinct to-the-point presentation of information is required.

Decision makers need them to deliver a comprehensive answer or response. Quite distinct from the point above, the information delivered must be holistic in nature. This will be a primary theme in subsequent sections of this book, but it is worth mentioning now that risks related to decisions emanate from a broad spectrum of disciplines including, for example, legal, environmental, commercial, financial, technical, and political. It is the decision-maker's nightmare to recommend a course of action only to find that a fatal, for example, legal foible was not considered and is of sufficient negative impact to invalidate the recommended decision.

The decision maker needs to believe them. Information should be delivered from the appropriate knowledgeable and credible sources. Such sources should be as free of bias as possible. In a large corporation this is a difficult achievement. In addition, to the extent possible, information should be delivered by individuals involved in the generation of the information to avoid massaging of the message.

The decision maker needs assurance. Regardless of relative rank, everyone experiences some level of self doubt and angst regarding important decisions. Almost above all, most decision makers value assurance from others involved in the process that the information upon which the decision is based and the decision itself is credible and valid. This assurance mainly comes through comprehensive implementation of the points listed above.

The decision maker is not looking for a better way to calculate something. Too many times, the proponents of a probabilistic process focus on how the calculations are made and the meaning of the numbers. This is all good stuff, but what the decision maker really wants is to be convinced that having such new information will actually impact the decision-making process and that the benefits of creating and considering such new information outweigh the costs. The onus is on the purveyor of the new process to be prepared to clearly demonstrate the link between creating and considering the proposed information and better decision making. This can be difficult to do, but it is an essential element. Subsequent sections of this book will focus on this aspect.

Now, the astute reader should be thinking: "What does this have to do in particular with risk-related decision making? Most of this stuff pertains to decision making in general, whether or not a risk/uncertainty process is involved."

And the reader would be right. However, the making and taking of decisions in the risk/uncertainty framework puts both positive and negative twists on the conventional decision-making process.

On the negative side, as previously mentioned, framing and making decisions in the risk/uncertainty and probabilistic world necessarily adds a level of complication to the generation and consumption of information. Successfully navigating in the probabilistic realm requires some level of competency regarding probabilistically-based data-collection and data-manipulation processes (the subject of most of the rest of this book) as well as the ability to comprehend and masterfully utilize probabilistic results such as probability curves, sensitivity analyses, and other outcomes. These are decision-maker traits that are not typical and certainly not universal.

On the positive side, considering information in a probabilistic mode enhances the information content. When a deterministic cost estimate is considered, for example, the decision maker rarely has any feel for whether this cost is easily achieved (near the high end of possibilities), is a most likely cost, or is a stretch goal target. When such cost estimates are considered as part of a range of possible outcomes, the probability of achieving that cost (or some greater or lesser cost) is clearly conveyed. Assurance is enhanced and more realistic plans can be set.

The bottom line for decision makers is that they want to know the following:

- What questions do I ask to determine the quality of the information being conveyed to me?
- How do I integrate the information to make good decisions?

All of the text in this section addresses, in a cursory way, these essential queries. However, the answer to these questions is more complex than has been related thus far. The following chapters address various aspects of these questions in what the author believes is a logical and convincing order. In a later chapter, after all arguments have been made, I will return specifically to these two questions to summarize the response to these questions posed in this book.

SELECTED READINGS

Barton, T. L., Shenkir, W. G., and Walker, P. L., *Making Enterprise Risk Management Pay Off: How Leading Companies Implement Risk Management*, Financial Times Prentice Hall, Englewood Cliffs, NJ, 2002.

Hiatt, J. M. and Creasey, T. J., *Change Management*, Prosci Research, Loveland, CO, 2003.

Pritchard, C. L., *Risk Management: Concepts and Guidance*, ESI International, Arlington, VA, 2001.

2 Risk Assessment: General Characteristics and Benefits

WHY RISK ASSESSMENT?

People are pressed into enacting risk assessments for various reasons. Foremost is the need to be able to consider a range of possibilities rather than a single-valued answer. Loosely speaking, a single-valued result is termed deterministic as opposed to a multi-valued (distribution-based) result that is referred to as probabilistic. The term probabilistic is used because, in addition to relating a range of possible values associated with the answer, a probabilistic result imparts the probability that values in given ranges will occur.

Another primary reason to perform a risk assessment is the uncertainty associated with many projects. For example, it may be that one of several properties is going to be purchased and the purchase price for each is fixed (no distribution of prices associated with each possible purchase). However, because of uncertainty associated with getting loans from lenders, the uncertainty that the properties will be ready for sale when you are ready to buy, the uncertainty related to the present owners being able to find other places to live, etc., you may want to deduce which investment is the most likely to succeed or fail. In this case, although you would be working with deterministic answers (fixed purchase prices), the probability or range of probabilities associated with each purchase may need to be known.

Risk assessment is also a very powerful front end for decision making and portfolio management. Although it is true that risk assessments sometimes are run on individual stand-alone situations, more commonly the aim of a risk assessment process is to assess several opportunities and to be able to compare, rank, and portfolio-manage the options. To do this, a robust risk model or models must be generated (models that can handle the varied aspects of all opportunities to be assessed), and a common measure must be agreed upon according to which all opportunities can be compared.

For example, a company may have x dollars to invest. There are several investment opportunities available. Two are purchase opportunities: one, a restaurant chain, and the other, several clothing stores. Two other opportunities are investment types. One investment is to provide start-up money for a new firm. The other is to invest in a particular international mutual fund. In this case, you would likely begin down the risk assessment road by deciding on what parameter you are going to use to compare these opportunities. Because they all involve the investment of dollars by the company, perhaps you would settle on determining the net present value (NPV) of each opportunity for comparison.

Having settled on NPV as the desired risk assessment result, it is likely that you would proceed to generate two models. One model might be designed to consider the aspects of purchasing another company. The model would need to be robust enough to be able to assess very different types of businesses — in this case, a clothing business and a restaurant business. This model, however, would generate as a result the NPV of each purchase. The second model would have to be comprehensive enough to handle different investment types. In this case, you are trying to compare a seed money commitment with a mutual fund investment. Again, like the business-purchase model, the desired output from the investment risk model is NPV.

Ultimately, these two models would result in four NPV distributions for comparison. The art (and sometimes science) of portfolio management is then applied to determine which opportunity or combination of opportunities will be pursued. In this way, risk assessment serves as an essential front end for decision making and portfolio management.

COMMON LANGUAGE

If I were to say the word cost, what would that mean to you? Do you think of the number on a price tag? Do you think of the price plus the tax? Do you imagine the dollar value of an item plus the cost of transportation to get to the place of purchase? Do you consider the value of your time? Are you thinking not just about the immediate cost, but of the total lifecycle cost of the item?

As I have alluded to already in this book, and will focus on in subsequent sections and chapters, risk assessment typically requires the participation of personnel representing disciplines that deal with law, the environment, commercial aspects, political concerns, technical problems, and so on. As the reader is likely well aware, each discipline creates and utilizes a jargon of its own, including unique terms that pertain exclusively to that discipline. Unique terms aside, different disciplines also share common terms, but, because of their unique perspectives, do not enjoy common definitions for the shared terms.

I have in the previous chapter defined the term risk. The definition that I offer is one that I have found can be embraced — sometimes reluctantly — by all disciplines. I was compelled to create this definition because as I interfaced with various groups that were in some way attempting to use risk-related techniques or processes, I quickly came to the realization that no two groups — and no two individuals within a group — shared a common definition of the term. How an individual perceives risk is, of course, colored by personal experiences, education, and other factors. So it is with all common terms.

It would seem a relatively simple thing to bring uniformity to the way people perceive, understand, and use terms. It is not a simple thing. At least part of a person's perception and understanding is emotionally based. Such psychologically entrenched views cannot be practically altered. So, if one is attempting to impose uniformity with regard to definitions of terms, can it be done? Well, yes and no.

It is folly to try to alter the personal perception of a term that is harbored by any individual. However, what can be done is to issue an agreed-upon set of definitions that everyone will use when interacting with one another. This is a different thing than the personal perception of a term. For example, if a person is an economist by training, they might tend to think of cost as the full lifecycle discounted cost of an entity. This definition might not jibe with the agreed-upon definition of the term. It is likely to be highly impractical to attempt to change the way the economist internally processes the term, but if there exists a glossary of terms in which the word cost is defined, then the economist can know that whenever he sees the term in a communiqué or hears the term in a conversation, the agreed-upon definition is what is meant.

There is no end to the confusion and wasted time and resources that can be directly and indirectly caused by misinterpretation of terms. Creation of a simple glossary of agreed-upon definitions can bring amazing clarity to the business process and avoid the bulk of unpleasant conversations that begin with, "Oh, I thought you meant...."

Because each company or part of a company will define differently any given common term, it is beyond the ken of this book to offer suggested definitions, except for those already offered for the terms risk and uncertainty. However, just a starting set of terms to be considered might include the following:

- Abject failure
- Consequence
- Deterministic estimate
- Event
- Expected value
- Failure
- Mitigation
- Probabilistic estimate
- Probability
- Risk
- Risk analysis
- Risk assessment
- Risk management
- Risk weighted
- Success
- Target
- Uncertainty
- Uncertainty assessment

This is by no means a comprehensive list. It is offered only to serve as an example of the types of terms that might be considered and defined. Now, some practical advice regarding the length of the list.

While the list should contain definitions of critical terms, it should not be so long or comprehensive that it unduly constrains communication. It is clear that the definition of certain terms such as risk needs to be universally agreed upon. If the list becomes too long, however, people will find that they cannot say or write anything without repeatedly referring to the definitions, the list of definitions being too much to carry in their heads. This is the short road to abandonment of the glossary. So, it is somewhat of an art to identify the terms that require agreed-upon definitions — and no more.

HOLISTIC THINKING

Creation of a cogent risk model — if that is ultimately required — begins with a comprehensive awareness of just what is the problem and just what needs to be known to resolve the issue. Subsequent chapters of this book are exclusively dedicated to these processes. Adopting the recommended techniques compels risk practitioners and their management to recognize the broad spectrum of conceptual, tangible, and probabilistic aspects and consequences of a given situation. That is, following the recommended risk assessment techniques is intended to promote holistic thinking.

Such a holistic cognitive process should impact multiple considerations:

- Scope of the project
- Techniques to be utilized
- Consequences
- Resources required
- Cost and benefits (aside from consequences listed above)
- Other aspects

Holistic thinking with regard to the scope of the project can be critical. Scope should include not only what the project will accomplish, but also what entities and disciplines will need to be involved. For example, the scope of a construction project should include not only the engineering and construction considerations, but also the legal, environmental, political, and other real-world project-related aspects. Often in the initial conversations and plans related to such a project, these seemingly ancillary attributes are not specifically considered as major impacts on project value. This is a mistake.

Most of the things that cause projects to fail — however we have defined success — are not line-items on spreadsheets. Causes of failure typically relate, for example, to political or legal problems that cause schedule slip, environmental problems that cause cost overruns, logistical impediments, and so on. A skilled risk-process facilitator will cause individuals or groups to consider and address quantitatively these softer issues, and to recognize their impacts and consequences.

Holistic thinking is also useful when considering the techniques and processes that might be applied in the course of project planning and execution. Following on from the points made above regarding holistic thinking related to scope, the multiple disciplines involved in assessing an opportunity will each bring unique methods, processes, and outputs to the assessment. Wasted time and money can be avoided if consideration is given to the range and disparate types of information that will need to be integrated to achieve a credible risk-weighted estimate of project value.

For example, finance department personnel might pose their analysis results as a risk/reward plot. Engineers might present their estimate of time as a Gant (Gantt) chart. Cost estimates from the engineers and economists could appear as cumulative frequency curves resulting from simulation techniques. A holistic view of the pending flood of disparate information types and formats can be critical to successful confluence of such information. The time to impart integration needs to various groups is not after they have started their processes or after they have generated output. Holistic thinking about the integration process and convening early meetings that focus on integration requirements can have major impacts on the amount of "do overs" associated with a project. What is the old saying? There is never enough time or money to do it right the first time, but there is always enough time and money to do it again.

AWARENESS

One of the greatest benefits to result from enacting a probabilistic assessment scheme in a corporation, yet one of the most vexing to attain, is that of general awareness within the organization. Typically, until people are forced to address the specific aspects of a project and are required to attempt to quantify impacts associated with project parameters — especially soft considerations such as political and cultural issues — awareness of real-world benefits and consequences is not fully realized. Adroit implementation of a probabilistic risk assessment process can result in the following:

- Awareness of resources required
- Awareness of just what is the problem
- Awareness of the time and cost (the *real* schedule)
- Awareness of training required
- Awareness of where the answer lies in a range of answers
- Awareness of consequences associated with individual parameters in an assessment and with the assessment process
- Awareness of probability of success
- Awareness of multiple measures of value and resulting consequences (two sets of books, etc.)
- Awareness that this will impact all levels in the company and how things are done
- Awareness that this process is difficult to overlay on the current business process

Each of the items listed above is covered in detail in another chapter in this book.

Of course, awareness alone is not sufficient to put into remission the ills that each of these points of awareness might address. In fact, awareness of some of these items might prompt undesirable and unintended (but, sometimes intended) consequences. For example, awareness of the realistic probability of success of a project can lead to questionable practices by those who have a stake in the project's success. However, without awareness of these and other aspects of project planning and execution, the right and necessary conversations between critical parties are much less likely to occur.

As stated at the very beginning of this book, the challenge is not primarily one of technical prowess. The salient challenge is related to changing behaviors. Until awareness is realized, critical parties meet, and the right conversations are fostered, little progress can be made toward realization of the practical and real-world value of a project.

CONSIDER RANGE OF POSSIBILITIES

Lo, those many years ago when I was a graduate student, as part of a geochemical experiment I would solve a system of equations that calculated a value. I hoped the calculated value would predict the outcome of the actual physical analysis. The single calculated value or series of values resulting from re-solving the equations with slightly different coefficients did not afford me any feel for whether my results would be even remotely coincident with real-world values. That is, the values resulting from re-solution of the equations with various coefficients did yield a range of answers, but I could never be sure how much of the actual (real-world) range this represented. In addition, it was just too time consuming to even attempt to compute all of the possible permutations.

A well-constructed risk model alleviates this conundrum by using ranges of values to represent most, if not all, variables that constitute a set of equations. Application of Monte Carlo techniques (see Chapter 11) to the variables and equations will yield a range of answers. The Monte Carlo technique will cause all reasonable combinations of variable coefficients to be combined in equations. This will yield an output curve that represents all probable solutions.

Given the sophistication of modern risk assessment software, there is no reason to resort to discretization of distributions (attempting to represent a distribution with a small number of discrete values). Reduction of a distribution to representative discrete values drastically reduces the information content of a variable and thus is to be avoided.

ESTABLISH PROBABILITIES

The aforementioned answer-variable distribution is nothing more than an array of x-axis values — usually sorted from smallest to largest (see Chapter 12). A risk assessment would not be a risk assessment without associating each x-axis value with its appropriate probability (see Output from Monte Carlo Analysis — The Frequency and Cumulative Frequency Plots in Chapter 11).

Each point plotted on a risk-model-output curve contains two pieces of information. One is the magnitude of the coefficient. For example, on a plot of profits, each point plotted would represent the number of dollars we could realize as profit. In addition, as part of a cumulative frequency plot, each point is assigned a probability that indicates the likelihood that you could attain that value or a value of greater or lesser magnitude.

Probabilities such as those described here, expressed as a cumulative frequency plot, aid significantly in the decision-making process. For example, if we need to realize a profit of $50MM from a venture and from the cumulative frequency plot we surmise that there is a 95% chance of realizing $50MM or more, it is likely that we will pursue the opportunity. However, if the plot indicates that there exists only a 45% chance of achieving our base-case profit, then the decision to go ahead with the project is not so cut and dried, so to speak. With a 45% chance of economic success, one would have to contemplate the avenues of gathering more information, weighing this opportunity against other opportunities, assessing the company's ability to sustain a loss if the opportunity fails, and other considerations to help us make our decision.

COMMUNICATION WITH CUSTOMERS AND THE PUBLIC

Risk models and the output from such models are potent vehicles for communicating with customers and the public. This communication avenue can be utilized in every business.

Consider the scenario in which you have been presented with an investment opportunity. The investment will require the outlay, on your part, of $100,000. Without a risk assessment, those attempting to entice you assemble a proposal that explains just how and over what time the return on investment will be generated. The pitch presents you with a deterministic estimate of the investment's return to you. The single value presented is $150,000 — a $50,000 profit. This appeals to you, and you decide to invest.

The investment is made, but because of bad weather, a short growing season, labor problems, and unforeseen transportation snafus, the investment returns only $100,500. You are outraged because any other competing investment would have had a superior return.

The scenario in which a risk assessment is applied is much different. Suppose the investment proponents engineer a risk model to estimate the range of possible returns. The model includes variables that force consideration and quantification of delays, the probability and consequence of labor problems, transportation availability, weather calamities, and so on. All of these parameters are represented as distributions.

The model terminates with a distribution of potential returns. The potential-return curve indicates that if there is a confluence of nearly all potential problems, there is a small but significant chance that nearly the entire initial investment of $100,000 could be lost. The most likely return is around $135,000, with a maximum, but unlikely, payback of almost $175,000.

Well, now you have to consider the distinct possibility that all of the investment could be lost and that there is a significant chance that the investment could return less than the initial $100,000 you might put in. The reasons for the possibility of loss have been explained to you variable by variable. After long consideration, you decide to go ahead with the investment.

Under this scenario, when the investment returns only $100,500, you are disappointed, to be sure, but are also relieved that the entire investment was not lost. In either scenario (deterministic or stochastic) the result is the same, but your reaction to the final value is drastically transformed by the application of risk assessment. See Chapter 6 for more information concerning risk and communication.

Consider now a lawyer who has used a risk model to calculate the potential damages for which his client may be responsible. Like the investment risk model described above, the legal model yields a range of potential liabilities. In the case of the investment risk assessment, communication centered mainly on the output variable representing return. Although the lawyer certainly would discuss in detail the output plot of expense, he would use the risk model primarily as a communication tool to foster discussion concerning the input variables for the model.

A comprehensive legal risk model typically contains a dozen or more input variables that are combined to yield the resultant liability distribution. Discussion of each of these variables and the range of values used to represent each variable is invaluable communication with the client. When the client understands and agrees to the range of the input values, the resultant distribution of potential liability is more easily accepted and most certainly is better understood.

PROBABILISTIC BRANCHING

It has already been mentioned that an output plot from a risk model conveys information regarding the probability of occurrence of a range of output values. However, probability can also be used to great advantage as an integral component of the risk model/equation logic.

Let's consider that we are in business with a partner. Our partner's business has fallen on hard times and it might go out of business in the coming year. One of the projects in our company depends upon the contribution of this partner. We would like to build a risk model to project the profits from our partnership project, but we realize that this risk model must consider the possibility of failure of the partner's business.

Meetings within our own company and with partner representatives lead us to believe that there is a 60% chance that the partner's business will survive the current financial downturn. Therefore, conversely, there is a 40% chance of economic failure for the project.

Our risk model, which calculates the net present value of the project, will then contain some probabilistic branching:

SUCCESS = 60
If Rand_num < SUCCESS then
    (Execute success plan)
Else
    (Execute failure plan)

where *Rand_num* is a randomly generated value between 0 and 100. The lines "Execute success plan" and "Execute failure plan" represent sets of equations that would calculate profit under each scenario. Our output plot of profit, therefore, would reflect the probability of our partner's failure. This is a simple example of a powerful and oft-used risk assessment technique.

RELATIONSHIPS BETWEEN VARIABLES

The real world is a complex place. Most occurrences in the real world have an effect on other happenings. If the price of food rises, it might delay the purchase of that new car. As demand for your product rises, the price you get for that product also can rise. As labor costs and taxes increase, your profit margin per unit might decrease.

Through the application of dependence (see Chapter 14), a properly constructed risk model can closely emulate real-world events. Let's consider the situation in which we would like to launch a new product in the marketplace. Three (of many) variables in our risk model are stage of technology (SOT), customer commitment (CC), and maximum cash exposure (MCE). From these and other variables, we would like to calculate the first-year net income for the potential new product. We believe that the technology lead on the competition is significant. However, it is not the actual technological edge that counts, but the customer's perception of our technological prowess that is paramount. Therefore, the distribution representing our stage of technology ranges from relatively low values to relatively high values because of uncertainty surrounding public perception.

We also believe that as our SOT is more favorably perceived by the public, our CC will increase. People like to go with the winner and prefer to buy from a company they believe will be in existence in the years to come. So, in our risk model, we establish a relationship between SOT and CC so that as SOT increases, so does CC.

As our SOT increases, it is easier and easier to use the money of investors, rather than our own cash, to fund the company. Investors are drawn to companies that are perceived to be leaders in their field. Therefore, as SOT increases, MCE (spending our own funds) decreases.

In this risk model, then, SOT is controlling the variables CC and MCE. If, on a given iteration in the Monte Carlo process (see Chapter 11 for more details), a high value of SOT is selected, the dependence mechanism will cause it, in this case, to be combined with a relatively high value for CC and with a relatively low value for MCE. Conversely, a low SOT value will be linked with a relatively low CC coefficient and a relatively high MCE number. Risk models that honor such relationships between variables can do an excellent job of emulating real-world processes.

LINKING OF DISCIPLINES

As alluded to previously, the real world is a complex place. A single opportunity generally is composed of many parts and represents multiple disciplines. A well-engineered risk model can logically and succinctly meld multiple disciplines into a single entity.

Consider the situation in which our environmental firm has been solicited to bid on a remediation project. The job would be to clean up surface and subsurface liquid waste. This will be an intricate, delicate, and costly venture. Our firm, of course, would only bid on the project if we thought we could profit from the effort.

To estimate the potential for profit from this project, we call together the heads of all departments. Our political analyst attends the meeting because we know from past experience that we likely will encounter political resistance in some shape or form. In the past, local opposition groups halted projects because they contended that proposed clean-up efforts fell short of what actually was required. We have also run into trouble with conflicting state, county, and local regulations. In addition, required permits often are slow in being issued at the state capital because of political wrangling.

Our scientists also attend the meeting because this project presents scientific challenges. We will have to be able to benchmark the level of contamination of the current situation and then monitor the toxin levels as we proceed with the cleanup. The offending substances can be toxic even in exceedingly minute quantities. Our scientists are not sure that we can make statistically substantiated claims that we have reduced the toxin concentration to the required level, given the low levels of concentration.

The meeting includes representatives from our engineering staff because there are myriad engineering challenges. Building special sampling equipment will be an especially daunting task. Also, we will need to drill many wells to monitor the subsurface toxins. We have to do this without disturbing the water table and without causing more surface pollution by the drilling process itself. Other engineering challenges include transportation and disposal of the toxic waste.

Representatives from our commercial arm attend the meeting because commercial aspects of the project also need to be considered. The cleanup effort will require the purchase of large quantities of special chemicals. There are only two companies that produce the desired commodities, and it is not clear whether we can obtain the necessary volumes in the time allotted.

Freight rates are astronomical for the transport of toxic waste by rail. The cleanup site is quite distant from the nearest rail line, so trucks also will need to be purchased for initial transport from the site.

Logistic, environmental, financial, and other considerations are also concerns. A risk model is a unique entity by which all aspects of a problem can be joined to result in a desired outcome (see Comprehensive Risk Assessment Example in Chapter 17 for details and examples). Prior to the use of risk assessment technology, such problems would be approached quite differently. Each department, usually without coordinating with other departments, would put forth their deterministic estimate. These individual values, then, would be aggregated to arrive at a single resultant value.

Results reached in such a manner rarely come close to the actual real-world value. Ignorance of the dynamics and interactions of the individual departments was alone enough to add significant error to the estimates. A well-constructed risk model, through the application of probability modeling, dependence, and other techniques, generally alleviates this malady. In addition, the iterative nature of Monte Carlo risk models attempts to ensure coverage of all reasonable permutations. Risk models, therefore, are a powerful and convenient means of bringing together and correctly integrating data from multiple disciplines.

CHECKLIST — DO NOT TAKE IT FOR GRANTED

One desirable attribute of a risk model is that it affords a list of variables. A user of a risk model is, in one respect, like an airline pilot. A pilot may have flown the London-to-Newcastle route 1000 times in the past. However, each time she climbs into the cockpit, she reviews the checklist of items that must be inspected or tested prior to the flight. So it is with risk-model users.

Consistent application of a risk model forces individuals to consider each and every element of an analysis and to not take anything for granted — no matter how many times the individual has previously executed the model. Many is the time I have seen that the sources of error in an analysis lie hidden in the items that were assumed or taken for granted.

Utilizing risk models forces not only acknowledgment of each variable, but fosters discussion concerning the range of values, etc., that will be used to represent that parameter. Even the most mundane and assumed items receive this treatment. This can be a critical practice.

For example, in the oil-well-drilling business there are parts of the world where there exists plenty of what is called source rock. These are the rocks that contain ancient plant and animal matter and, when deeply buried, produce oil and gas from the organic material. The oil and gas produced in these rocks migrates upward through the rocks of the earth's crust until, hopefully, it is captured and held by any number of geological mechanisms. We call these places of capture traps.

An oil-and-gas risk assessment model typically contains subsets. Each subset of the model addresses some aspect of the petroleum system. Just a few of these aspects are the richness of the source rock, its volume, depth of burial, geothermal gradient, migration pathways, trap-sealing mechanism, porosity, permeability, water saturation, trap volume, recovery factor, and formation volume factor.

In the special areas of the world where it was traditionally assumed that shortage of source rock was not a problem, users of an oil-prospect risk-assessment model typically turn off the source-rock section of the model. They might do this by supplying default coefficients that would generate far more oil than could be held in any single trap. Later, problems might arise regarding matching oil reserves from less than half-full, real-world traps with the reserves estimated by the risk model. Explorationists, for a while, might be stumped. The problem might lie in the assumed variables. In fact the traps being tested might not have been originally filled with oil because of a local lack of source rock and deficient migration pathways. The lesson learned in this case would be to use the list of risk-model variables as a checklist and to consider each variable for each execution of the model. Risk models, therefore, can serve as an excellent means of enforcing completeness and consistency in any evaluation process.

UPDATING ESTIMATES

The world is a dynamic place. Things change all the time. As soon as you complete an exhaustive and comprehensive evaluation of a situation, it is out of date and practically useless. Such is life.

Risk models afford us the means to at least attempt to keep pace with the ever-changing world. A properly designed risk model is not just a set of probabilistic equations, but is a user interface by which users can nearly effortlessly and without having to supply statistical parameters (mean, standard deviation, etc.) enter data into the system and cause it to execute. A fully integrated risk model also should serve as but one link in a process chain. That is, a complete evaluation process should involve a data gathering and verification step (see Users Will Game the System — The Risk Police in Chapter 5), a data processing function that perhaps includes a risk assessment, and a portfolio-management function. In such comprehensive processes, it is not uncommon for the user interface of the risk model to serve as the primary user interface for the entire system. That is, all data are either passed to the risk model from data sources such as databases, or are entered directly into the risk-model user interface.

In such a configuration, a user can approach the risk-model interface and call up the previous data entered into the risk model. If the changing world has caused a portion of that data to become obsolete, a user can utilize the risk-model interface to quickly and easily supplant the obsolete entries with relevant data and re-execute the entire process. This process can, in a properly designed system, result in refreshed risk-model output and a new mix of opportunities resulting from the linked portfolio-management process.

In this way and others, an integrated risk assessment system can afford a quick and nearly painless means for updating databases and for making business decisions. The real power of a risk model only is realized when opportunities of similar ilk are consistently assessed and when the risk model is executed as part of a process that results in portfolio management.

As an example of this process, let's consider a brokerage house that maintains a portfolio of investments for clients. This portfolio is designed to generate the highest return on investment at a given risk level. The portfolio is dynamic because market conditions, portfolio-member companies, and myriad other things change on a minute-by-minute basis.

Firm executives have decided that they would like the ability to update the portfolio on a daily basis. Each day, data are automatically collected and stored in the firm's database. In addition, the firm's financial analysts execute other systems and use their market knowledge and experience to daily generate forecasts, trends, and other information.

Each day, an individual at the firm can approach the risk-model user interface and, using the mouse, double click on a risk-system icon that causes the risk model to be populated with the latest information stored in the firm's database. The interface also affords the user a quick and easy means to enter other forecast and analytical data generated by the analysts. When all data have been entered, a click on another icon executes the risk model. Probabilistic time-series and other stochastic data are automatically passed from the risk model to the portfolio-management routine that uses portfolio-management technology to automatically calculate a new mix of investments. With a properly designed risk model, this process is economically and technically achievable. This process can be a powerful adjunct to almost any business.

TO WHAT TYPES OF PROBLEMS CAN RISK ASSESSMENT BE APPLIED?

Risk assessment is a universal and generic tool. Typical model types include health and safety (fate/transport type), environmental, transportation, construction, exploration, legal, insurance, and many other types. To impart to the reader a sense of the breadth of application of this tool, I have listed below summaries of just a very few of the models upon which I have worked.

LEGAL MODEL

A model that calculated the net benefit of settlement vs. litigation was built to aid in legal decisions. Net benefit of settlement, net cost of litigation, net cost of settlement, total cost of verdict, and other outputs were calculated. Input variables that comprised broad categories included parameters such as litigation costs, total damages, likelihood of a verdict, and other probabilistic and economic aspects.

ENVIRONMENTAL HEALTH AND SAFETY MODEL

Models were built that calculated the total environmental, health, and safety risk and cost associated with entry by the company into various countries around the world. Risk of subcomponents of the model also were calculated and presented. Input-variable categories (many variables per category) included public perception considerations, government approvals and permits, ecological and cultural parameters, health and safety considerations, and evaluation of preexisting damage.

PIPELINE ROUTE SELECTION MODEL

A comprehensive time-series model was constructed to help a consortium decide which of several routes would be selected to construct a pipeline for a major oil field. The pipeline route-selection model calculated tariffs and other parameters from variables that represented major categories of consideration. These included political concerns, environmental problems, commercial considerations, financial parameters, technical considerations, taxes (for many countries), and other parameters. This model was used with great success to rank and prioritize routes.

POLITICAL MODELS

Models were constructed that evaluated other countries based on categories of variables. Categories included political stability, foreign investment conditions, operating environment, transportation infrastructure, and other considerations. Results from the models facilitated comparison of countries on a common scale.

CAPITAL PROJECT RANKING AND PORTFOLIO MANAGEMENT MODEL

This model calculated profitability index (PI), internal rate of return (IRR), net present value (NPV), and other financial outputs. Input variables included project safety and environmental aspects, cost estimates, incentives, discount rates, taxes, maintenance and insurance costs, and other considerations. This model was run on all capital projects at a manufacturing facility. The projects were ranked and portfolio-managed based upon the model outputs.

NEW PRODUCTS MODEL

Research and development organizations generate products and processes, each of which may have commercial value. It is, however, expensive to launch a new product in the marketplace. Therefore, products and processes need to be ranked. A model was built that included categories of variables. Some of the categories were technical considerations, marketing aspects, financial and commercial facets, and others. The model facilitated the application of weights to the various considerations. Results from the model allowed the prioritization of potential new products.

FATE/TRANSPORT MODEL

A model was constructed to calculate inhalation exposure. Exposure was represented in the model as the average daily dose for noncarcinogens and the lifetime average daily dose for carcinogens. Among the input variables were parameters such as concentration of chemicals in the air, inhalation rate, bioavailability, exposure duration, exposure frequency, body weight, average lifetime, and other considerations.

SELECTED READINGS

Finkler, S. A., *Finance and Accounting for Nonfinancial Managers*, Prentice Hall, Englewood Cliffs, NJ, 1992.

Newendorp, P. D. and Schuyler, J. R., *Decision Analysis for Petroleum Exploration, 2nd Edition*, Planning Press, Aurora, CO, 2000.

Nijhuis, H. J. and Baak, A. B., A calibrated prospect appraisal system, *Proceedings of Indonesian Petroleum Association* (IPA 90-222), Nineteenth Annual Convention, 69–83, October, 1990.

Rodricks, J. V., *Calculated Risks — Understanding the Toxicology and Human Health Risks of Chemicals in Our Environment*, Cambridge University Press, Cambridge, MA, 1994.

3 Risk Assessment — A Process

RISK ASSESSMENT — THE PROCESS

The marketplace abounds with risk-assessment software and consultants who are experts at applying risk-related systems to the problems of their clients. Such applications tend to be relatively sedulous academic and technical exercises. In addition, these applications generally are aimed at the assessment of risk associated with a single project or a small number of opportunities.

In the nearly 20 years I have been involved in statistical analysis and risk assessment, if there is one thing I have learned (and it took me quite a number of years to learn it because, in the beginning of my career, I knew everything), it is that risk assessment is a process. The technical aspects of software selection and risk-model building are relatively trivial when compared to the ergonomic, organizational, political, communication, and educational aspects of implementing risk assessment in an organization. Technical brilliance alone does not come close to being sufficient.

To begin with, risk assessments rarely yield maximum benefit when applied in one-off or single-project situations. Risk assessment and risk models should be viewed as a means of consistently assessing multiple opportunities or projects with the aim of comparing, ranking, and portfolio-managing the independently assessed entities.

The risk process begins with two fundamental elements: a need (usually ill-defined) on the part of an individual or group and a vision held by the person or cadre of bodies that will implement the stochastic solution. A call for help usually is the catalyst.

As detailed in subsequent chapters, the facilitator of the risk process, upon answering the call for help, must cajole, prod, and sometimes trick those in need into realizing and expressing just what really is the problem. This is a nebulous and contentious undertaking, even (sometimes especially) when those in need are doing their utmost to cooperate. If you are contemplating the possibility of acting as a risk-process facilitator and are troubled by confrontation and piqued by chaos, then you should stop now.

Shepherding a group through the process of defining the problem and its component parts is not for the faint of heart or for those with a paucity of will. A blend of interpersonal skills, leadership, patience, and technical competence are requisite. It is paramount for the facilitator to ferret out the fundamental dilemma and to lead the group toward a resolution that is practical and succinct, and that has been reached by consensus.

To be able to enact this chain of events, the facilitator must from the beginning have a vision of where this rambling train should terminate. That is not to imply that the facilitator should be capable of envisioning the answer — typically the facilitator knows less than anyone else involved about the troubled business. However, in a competent facilitator's mind it should crystallize early in the process just which parties need to be brought together, which individuals or groups are likely to be and remain detractors, what disciplines will have to be melded (i.e., legal, environmental, financial, etc.), and whether the perceived problem is a risk problem at all.

Asking pertinent questions in proper formats, guarding against creating too detailed or too general a model, being cognizant of attempts at double dipping (accounting for the same effect more than once), integrating chance of failure, enacting dependence when required, engineering a comprehensive and succinct set of equations, and other tasks are the primary responsibilities of the risk-model facilitator. Although actually important to the overall process of building and implementing a successful risk assessment process, the aforementioned tasks are primarily procedural or mechanical and represent but the proverbial tip of the iceberg.

Organizational consequences have to be considered. Foisting a risk assessment process or model upon an organization will not only change how opportunities or liabilities are assessed, but will significantly alter the way an organization makes critical decisions. As detailed later in this book, such decision-making processes typically lead to selection of winners and losers. Although these two classes might have evolved under any selection scheme, the risk model is an easy and popular source of and target for the angst felt by those who have been "done wrong." Change generally is distasteful. A strong commitment by the organization to the new and different quantitative and stochastic lifestyle is essential.

Other organizational upheavals are necessary as well. If multiple opportunities are to be assessed, ranked, and portfolio-managed, then it is critical that a common and consensus-built risk model be used to achieve consistency in assessment. Following the launch of a risk process and model, it is not long before those using the model get wise to what needs to be put in the front end to receive a favorable result on the back end. The only real guard against gaming the system is the imposition of a board of review. Members of the board are charged with reviewing all inputs to the system. Organizationally, this is tantamount to creating Big Brother. Without careful and delicate implementation, this organizational step can consume all the energy of those attempting to promote this necessary aspect of the risk process.

As you might guess, politics also can raise its ugly head. Just wait until a pet project of a powerful political entity is found by the risk process to be lacking relative to other opportunities and, therefore, is passed over for execution. Staunch supporters can change their stripes when they feel betrayed by the process they helped to create. Genuine political support and an organizational means of consoling those whose projects are not selected should be part of any worthy risk-implementation plan.

Instituting methods of consoling those associated with projects not selected is just one cultural adjustment that must accompany an institutionalized risk process. In some corporate cultures, salesmanship and the good old boys system are critical ingredients in the recipe for project success. With the advent of an impartial and quantitative risk process (review boards and all), parameters that in the past were ignored or taken for granted now are quantified, scrutinized, and invariably integrated to achieve a consistent result. Ranking of projects can be a much more businesslike and bottom-line oriented affair. This is a drastic cultural change in many organizations.

An equally difficult adjustment can be the newfound ability to call the baby ugly. In many businesses, individuals have for years put forward proposals that had a high rate of economic failure, or were not as attractive an economic venture as passed-over competing projects. Those individuals who had previously proposed such less-than-optimal projects will, with the advent of a consistent risk process, have to confront the fact that their baby, like many in the past, is pretty ugly. This can be a wrenching cultural adjustment and should not be underestimated.

Still another salient aspect of risk-process implementation is that of education. Obviously, those individuals and groups charged with feeding and executing the risk model need to be educated with regard to the technical and mechanical nuances of the risk model. This type of education is time-consuming, but pedestrian.

The ultimate educational challenge rests in bringing understanding of the output of risk models to those who will (or should) use the stochastic data in their decision-making process. Often I have witnessed the successful implementation of a risk process with the troops, only to have the process die on the vine due to the inability of decision makers to appreciate, understand, and utilize risk-model output. Systems and processes must be put in place not only to generate risk assessments, but to facilitate the use of risk-model output by the organization. A large part of creating this organizational capability is educational in nature.

The preceding paragraphs outline just a few of the components of a well-constructed and -implemented risk process. While it is true enough that not all (or even most) risk processes and models become formally institutionalized in organizations, each process should be approached as though it will be thus incorporated.

It has been my experience that such models can become very successful parts of a relatively small arm of an organization. Competing or simply related organizational entities take note of the acclaim directed toward the much more efficient risk-process-embracing group. Before you know it, hastily constructed and poorly implemented risk processes are rampant and, in most instances, do more harm than good. Poorly implemented risk processes are likely to be worse than none at all.

THE TIME IT TAKES TO DO IT RIGHT, AND WHEN TO GET STARTED

Throughout this book I will be referring to processes and facilitation techniques that form the core of the risk assessment exercise. These processes and techniques are sometimes iterative and time-consuming, and require the repeated participation of individuals and groups deemed critical to the solution of the problem. The belief that risk assessment is a last step is one that is widely held but nonetheless misguided.

As will be described in detail in later chapters, the process of generating a comprehensive risk model should spearhead any effort. The risk-model process outlined will do the following:

- Cause the critical parties to meet
- Foster discussion of the problem and its components
- Focus the entire effort on just what is the problem and just what is required as a solution
- Provide a basis for categorizing the parts of the problem and identifying critical variables
- Promote discussion of the various aspects of the problem and how they will be represented and combined to arrive at the desired solution
- Provide a means to delegate and distribute problem-related tasks
- In some cases, short-circuit the entire effort if the parties realize that perhaps it should not be done

The process of preparing a valid risk model, therefore, should be one of the first tasks undertaken in a project. This task requires the participation of people — sometimes many people. In turn, this requires the skills and talents of a practiced group/process facilitator. Leading a group through the process of building a risk model is, as the old saying goes, a bit like herding cats. Risk assessments can, and usually do, involve many aspects of a business including financial, commercial, technical, environmental, political, and others. Participants from each of these disciplines (and perhaps many more) may be necessary to the solution of the problem. Opinions on what is important, expectations for the result, methodologies of approach to problems, patience levels, turf considerations, and many other human-related aspects need to be considered and deftly handled.

Working with a variety of participants and working through the process toward the solution of the problem takes time. A correctly designed and handled risk assessment process is a thing of beauty. From the start of a project, a well-orchestrated process can focus the effort and participants on a problem that enjoys consensus and on a solution that is clearly defined. Responsibilities of individual groups should also be identified.

WHAT IS TYPICALLY DONE

There are some things at which quintessential corporate personnel are relatively skilled. Conducting meetings is one. Generating reports and lists is another. Assigning responsibilities is yet another.

In most corporations, groups might meet to discuss risks and uncertainties related to a project. From these meetings, lists of risks and uncertainties can be generated and responsibilities to mitigate risks and reduce uncertainties assigned. This, however, does not integrate the impacts of the risks (correlated or not) and does not necessarily quantitatively and consistently impact the value – however measured – of the project.

A typical corporate structure is both horizontally and vertically segmented. Horizontal segmentation is represented by division of the corporation into departments or functional entities. Departments such as finance, security, health and safety, commercial, and marketing are examples. Within each of these departments, and in the corporate structure as a whole, vertical segregation is also common. Within a department there are, for example, business unit personnel, supervisors for those people, managers who oversee a group of supervisors, and so on. So, why am I taking up space here conveying something you already know? The answer is, because the typical corporate structure has a profound impact on what is usually done about risk assessment.

I have already touched on, and will discuss in more detail in a subsequent chapter, the problem related to integration of risk assessments. In many corporations, individual departments, for example, act as autonomous entities. If, and that is a big if, these departments actually perform any sort of organized and practical risk assessment, the metrics measured and the format of the assessment output rarely are designed to dovetail with assessments performed in other departments. A finance department, for example, might produce a risk/reward plot. The assessment process employed by the commercial department folks might result in a risk matrix plot. Although these two approaches are valid and might be the best fit-for-purpose assessment types for these departments, how the metrics resulting from both assessments can be integrated to impact the perceived value of a project is not apparent or is at least a nontrivial exercise. This type of coalescence conundrum is typical of a horizontally segmented corporation. As will be addressed in detail in a subsequent chapter, integration of these individual assessments is critical.

The paragraph above addresses just one of the risk-process-related hurdles that needs to be overcome in a typical horizontally-segmented corporation. So, what about vertical divisions? Risk and uncertainty assessments typically begin with the folks in the trenches. That is, it usually is the group of engineers or scientists or others at the pointy end of the spear who generate the initial analyses. After all, it is these people who have the best view of real-world risks associated with project planning and execution. Their view of political problems, competitors, and other more global situations might not be as acute as those individuals whose job it is to oversee projects at a higher level in the corporation, but ground-floor assessments of project-related problems usually are generated by those most intimately involved in project planning and with in-the-field execution.

What Is Typically Done is the heading for this chapter segment. So, with regard to vertically-divided businesses, what is typically done is that assessments generated at lower levels are passed up the management chain for review. How many layers of management hovering over the low levels varies — some organizations are flatter than others — but there nearly always exists at least one management level between the low levels and the top of the house.

People who occupy positions in these intermediate levels might augment a received risk assessment in several ways. First, as already stated, people in these posts might be better positioned to add to the assessment higher level or softer risk and uncertainty impacts that are not usually considered by lower-level personnel. Such things as competitor analysis, market swings and penetration points, political concerns, and other acknowledgments usually are the purview of these folks. Addition of the impact of such considerations on the value of the project usually is performed at this level. This is, for the most part, a positive development (even though it might have a negative impact on the perceived value of the project).

If this process represented the sum of all that transpired at subsequent vertical levels, the world would be a rosy place. However, something else that is a bit more diabolical can take place in these multiple and ever-higher-in-the-corporation risk reviews. It is not uncommon for people at higher levels to decide that the project was too harshly assessed at the lower levels. The result of such assessment modification can be that the business unit can do more with less. It is not uncommon that assessments that reach the top of the house do not faithfully reflect the views and opinions of those who initially generated the assessment at lower levels. This massaging of the numbers in subsequent reviews is a trait of most businesses — regardless of the type of business. There are several organizational cures for this malady which will be discussed in subsequent sections of this book. It is enough, here, to indicate that this type of behavior is not uncommon and should be expected.

WHAT IS NOT TYPICALLY DONE

It has already been indicated, in the preceding section, that departments within a corporation — if they do perform risk/uncertainty assessments — typically employ a fit-for-purpose analysis technique. One thing that is not typically done is to give sufficient consideration to how this department-specific assessment process and output might need to be integrated with other such assessment results. If disparate assessment metrics, formats, and so on from individual departments are gathered and reviewed by a skilled risk-assessment facilitator, through guided conversations and a lot of hard work the different measurements can be integrated such that they all impact, for example, the projected cash flows and economic measurements (such as NPV, IRR, etc.) for a project. This is a nontrivial task. It is rare that a corporation employs a sufficient number of such skilled facilitators to execute this process on all projects. Therefore, one thing that is typically not done is the true integration of all pertinent risks and uncertainties to impact perceived project value.

Just one other thing that typically is not done relates to awareness. As delineated in the Awareness section in Chapter 2, people who are prompted to employ risk-related techniques, or those to whom results from such techniques will be delivered, can generally be unaware of the costs, work, and other ramifications for their business that stem from implementation of risk-related processes. Typically what is not done is sufficient preparation of the organization regarding these consequences. Implementation failure can result primarily from surprise. People can exhibit the knee-jerk reaction of rejection of a process not so much because the consequence to them is so great, but because they were taken by surprise (and sometimes embarrassed) by the ramifications of process implementation. Through adequate communication, the element of surprise can be reduced or eradicated. This, however, is rarely done or done well.

A final salient point I will address in this section relates to the big picture. When consideration is given to implementation of a risk/uncertainty process in a business, it should be somebody's job to look at the big picture and attempt to predict how this will impact the entire organization. I have already pointed out that some consideration should be given, for example, to how risk assessments from finance, engineering, safety and other departments need to be brought together to impact the perceived value of a project. This is but one aspect of looking at the big picture.

Implementation of a risk-related process will impact how budgets are allocated. Some consideration should be given to how budgeting will be handled in a probabilistic process (more on this later). It will have consequences for how different departments will interact with one another. Physical mechanisms for departmental interaction should be established. Some consideration needs to be given to the generation of corporate-wide guidelines regarding just what a risk/uncertainty assessment will measure and how this information will be conveyed to higher levels in the corporation. Preparation of upper management to receive probabilistic data and to be able to use such data in decision making is critical.

Preparation of the organization regarding all of these things — and more — is essential if the implementation process is to yield optimal results. These things, however, are easily overlooked and are typically not adequately addressed.

WHAT TO PROVIDE

The old adage "The customer is always right" should be tempered with "Well, sometimes." When it comes to performing a risk analysis, there is no question that the results of the effort should be a model that brings joy and contentment to the hearts of the customers, but rarely is it true that the customers have in mind a practical approach.

The quintessential example of this is nearly any risk model built for engineers. Let me preface this by saying that some of my best friends are engineers. However, when collaborating on a risk model with engineers, it always seems that there is no detail too minute to warrant exclusion from the model. Engineers are not alone in displaying this characteristic. Most folks who solicit help in building a stochastic model are much too close to the problem and believe that the veritable cornucopia of complications with which they daily contend are relevant. If you in fact built the risk model they asked for, it would undoubtedly, even for them, prove too large and unwieldy to be practically applied.

In subsequent chapters of this book are detailed the processes that should be applied to ferret out just what is the conundrum to be solved and precisely the methodology that should be used to solve it. I will here preempt the following chapters a bit by stating that the practiced risk-model builder should apply good listening skills and powers of reason and deduction in an effort to state, as simply as possible, just what problem it is that can be practically tackled.

No doubt, details are important, and are purported to be where the devil resides. However, the adages "Time is of the essence" and "Simple is always best" should be guiding lights. It should always be the aim to build a model that is as high-level as it is practical. For example, you would not build a construction model in which is generated a separate variable for each individual rivet, piece of steel, length of pipe, and so on. On the other hand, it is likely an oversimplification simply to concoct a single variable called costs. It is a tricky business to build a high-level model that contains sufficient detail to make people happy. High-level models have the advantage of being simpler to feed, simpler to run, simpler to maintain, and more universally applicable.

Usually, though, there comes a time when every high-level model is modified to handle a changing situation. There are two basic philosophies to the art of model expansion. One is to add on to the model to accommodate the new situations. The other is to raise the model to an even higher level.

For example, we may have a model that evaluates blue cars and green cars. Next thing you know, someone comes along with a red car and says, "Hey, your model cannot handle this." Well what do you do? Using the make it bigger philosophy, you would add a new red car variable to the model. Alternatively, using the make it higher principle, you might discard the original blue car and green car variables for two new variables called car and color. Now you can handle any car of any color without resorting to expansion of the model.

Next, however, someone comes along with a light truck and points out that your model is again deficient. The make it bigger folks add a light truck variable to their green car, blue car, and red car model. (Where does this end?) Many spreadsheets (you know the ones I mean) grow like this over time. The make it higher people again attempt to produce an all-encompassing model by replacing the car variable with the variable four-wheeled, gas-powered conveyances weighing less than 4000 pounds. So they now have a model composed of this new variable and color which handles all the situations they have seen thus far.

Obviously (or at least it should be obvious), both these model-expansion philosophies can be carried to ridiculous extreme. The model builder's task is to balance these two philosophies so that the model can indeed handle a wide variety of situations and yet be simple enough so that it can be practically applied. Again, subsequent chapters outline the details of this process.

There still are other situations in which customer demands have to be taken with the proverbial grain of salt. It happens, figuratively speaking, that you will get a call for help from the guy responsible for arranging deck chairs on the sinking ship. He believes he has a real problem with the chairs because they keep sliding down a deck of ever-increasing pitch. You, however, taking the broader view, astutely deduce that the problem actually lies elsewhere. This somewhat macabre scenario has its real-life corollary.

Consider the situation in which you have been called to help the waterfront and storage department become more efficient and save some money. Their perceived problem is that they would like to increase dock space for ships and increase onshore storage. They believe this is necessary because onshore storage tanks are full, so the ship that delivers raw materials cannot be unloaded. The ship, therefore, hogs valuable dock space for an inordinate amount of time. The individuals that oversee the dock and onshore storage believe they would like you to aid them in the construction of a risk model that will weigh the relative merits of several competing dock- and storage-tank-expansion plans. Because of your past experience with similar projects, you immediately deduce that this is not a construction or expansion problem, but is one of optimization.

The ship is large and, therefore, can deliver vast quantities of raw materials. The ship is fast and can make a round trip from the raw-materials-production site to the dock in question much more quickly than the raw materials can be consumed by the manufacturing process. The ship and crew sit idle for inordinate lengths of time because they cannot unload the cargo. Although the dock manager who requested your services takes the parochial view that this is a dock/storage problem, you can see that the source of the dilemma is ineptitude in shipping optimization. The guy who hired you has real and immediate problems and does not wish to discuss the relative merits of efficient shipping schedules. You deem it an incredible waste of time and money to go ahead with the very expensive dock and storage expansion plan. What do you do?

Given that the title of this section is What to Provide, the first thing to do is not to provide the dock manager with the model he requested. Prior to communicating your refusal, however, you should put together at least a rudimentary model that, in your opinion, could solve the problem. It clearly is a waste of potentially productive dock space to have it occupied by a ship that cannot unload. Even worse is the underutilization of the ship itself and the payments to crew members who are, essentially, idle.

The ship is too fast and of too great a cargo capacity for the job. You might hypothesize the use of two barges in place of the ship. Compared to the ship, barges are cheaper to buy, operate, maintain, and crew. Barges also can be of significantly smaller capacity and are slower. If sailing schedules for the barges are efficiently synchronized, the dock manager's problem would be nonexistent. With the use of smaller, slower barges, everybody wins. No funds are used to modify the docks or onshore storage facility. The dock space is used only long enough to unload the barges. Barges can always unload immediately because they carry less cargo and are correctly scheduled. The expense of the ship and crew is, relative to the cost and crewing of barges, greatly reduced.

A risk model should be constructed to emulate the barge scenario and to empha- size the aforementioned advantages. When risk-model results are presented to the dock manager, it should be in a meeting that includes a representative from the shipping department who is empowered to recommend action on the barge scenario. If a shipping representative is not present when the potential solution is presented to the dock manager, the dock manager is likely to reject the proposal. His rejection is probable because the proposal recommends solutions that he has no power to enact (under the proposed plan, taking the heat off him depends on someone else doing something) and because it does not, in his view, solve his immediate problem. If a shipping representative buys into the plan, then clearly the dock-expansion proposal is a waste of money and effort, and the dock manager is relieved.

So, what is the point of this maritime diatribe? The point is, do not always be quick to provide just what the customer asks for. I consider it my job not only to provide risk-related solutions to problems, but to also provide the best solution. It is often the case that those seeking help are much too close to the problem, so to speak, and the problem they present either is not the real problem or it is only a small part of a much more extensive malady. Enacting this approach, as with the hypothetical dock manager, is an interpersonal skill and organizational problem first, and a risk problem second.

At least one more fundamental scenario exists in which honoring a customer's request for a risk assessment and model would constitute providing the requester with a less-than-optimal response. Because application of risk technology and processes can yield very positive consequences for those who embrace them, entities who observe the success of others can misinterpret the application of risk techniques as a problem-solving panacea. It occasionally happens that a risk assessment is requested to resolve an issue that could not possibly be mitigated by application of risk-related technology.

Consider the situation in which you have been called upon to apply risk problem-solving techniques to purchase-order processing. The requesting group is dismayed at the organization-wide time and expense associated with the processing of purchase orders. They are considering the purchase prices and implementation costs associated with several competing vendor-supplied order-processing systems. The selected system would be installed at the order-processing headquarters. Your job, as they see it, would be to build a risk model that would compare the long-term benefits (cash flow over time and NPV) of the competing vendor systems so they can base their selection on a quantitative and probabilistic comparison.

You have worked in various parts of the company in which purchase orders originate. It is clear to you, after preliminary discussions with the group, that the real time-related problems and expenses associated with order processing cannot be solved by the implementation of a slick order-processing system at headquarters. Rather, the real wrenches in the order-processing works are organizational in nature, having to do with purchase-location approval chains and purchase-location/head-quarters interactions.

Certainly it would be perceived as presumptuous, arrogant, and a bit insulting if you suggest that they are barking up the wrong tree altogether and that a risk assessment model really would be of little or no benefit. The task at hand, then, is to cause the group to come to the realization that the remedy for the real problem is not a risk model. Well, how do you do that?

Elsewhere in this book I describe in detail the process by which a group comes to consensus concerning just what is the question to be addressed. As the risk-model facilitator, you must lead the group through a process that includes the following steps:

- Brainstorm and record possible statements of the problem
- Use an iterative voting process to sort through the brainstorming items and select the one(s) upon which a risk model would be focused
- Build a contributing-factor plot for the problem on which the answer and all contributing factors are plotted

As a stealthy and wily facilitator, you should in the brainstorming and/or sorting sessions be able to guide the group to state just what is the actual problem. They may believe at the session inception that the question to be answered has to do with comparing long-term economics of competing vendor systems. However, you should challenge the group to examine the problem in greater depth and guide them toward the mutual revelation that the problem they originally asked to be solved is not the problem at all.

If the process gets as far as beginning to build the contributing-factor diagram and they still have not realized just what is the problem, you might have to delicately suggest it yourself (the result of a sudden flash of inspiration, of course). In this case, you should make every effort to get them to understand just what is the actual problem. You should not provide them with the risk assessment or model that was the focus of their original request.

SIZE DOES NOT MATTER (NOT THAT I WAS WORRIED ABOUT THAT)

When I first started out in this risk business, I was usually proportionally intimidated by the size of the organization I was to address. Bigger usually equates to more people. More people usually equates, in turn, to a more complicated organizational structure and more complex interactions between entities. I believed, in the beginning, that layers of organizational obfuscation would make my job harder. While it is still true that it takes more time to impact a larger organization, I am no longer of the opinion that it is necessarily more difficult to implement a risk/uncertainty process in larger entities. My opinion has changed due to one thing — leadership.

Again, in the beginning, the tack I would take was to penetrate the organization at lower levels. I took this approach for several reasons. First, lower-level people were the ones I could most easily approach and whose time I could most readily usurp. In addition, it seemed obvious to me that this is where the action is and where risk assessments would likely begin. Also, early in my career, these were the people with whom I had the most in common, so I felt relatively comfortable interacting with front line folks who were actually executing the project.

Over time, however, I realized that this bottom-up approach was not the best practice. I remember my days as a young scientist at a research facility. Email was just being introduced. (Yes, there was a time before email.) Tech-heads like myself jumped right on the bandwagon. Some more senior management, however, were resistant to the change. To resolve the problem, the V.P. of research one day issued an edict. A letter circulated indicating that if you wanted to talk to him, the only way you could now get his attention was through an email message. Other attempts at communication would, for now, be rebuffed. Needless to say, his direct reports and others suddenly decided that this email thing might not be so bad.

My memory of this episode served to change my tactics when approaching an organization. I now recommend that purveyors of risk processes first attempt to convince the organization's top echelon that adopting risk/uncertainty-related processes is in the best interest of their business. On their part, a willingness to listen to (but not always buy into) such a pitch is enhanced when the purveyor has an established track record and is himself a relatively senior organizational member.

It needs to be stressed in such conversations with senior management that an edict to implement a risk/uncertainty process should not be issued prior to designing a roll-out process and securing the necessary resources to begin to imbed the process in the organization. Believe me, having no process is better than having a poorly-implemented one.

This top-down approach sports a couple of advantages. First, it is usually an afterthought in the implementation process to prepare upper management to actually understand and use the risk-related (i.e., probabilistic) data in their decision-making process. If upper management are initially accosted and the approach includes at least some introduction to how such probabilistic information has successfully been used elsewhere, it is typical that top managers will be willing to be trained in the use of such information.

Another benefit to the top-down approach is, I guess, the obvious one. When senior management has been convinced that the risk-related approach is the way to go and they express this to the organization, take-up of the processes at lower levels is much enhanced. (This, however, does not mean that lower-level folks will like it — or you — any more than they otherwise would.)

So, it is my considered opinion that the size of the organization is nearly irrelevant if the approach to implementation is carefully considered. The top-down method, while not without its problems, is generally recommended. Later in this book I will point out that in order to impress the upper echelon, real examples of success should be utilized. These examples, of course, are generated at lower levels in the organization. So, practical application has to first be proved at lower levels but spreading the practice throughout the organization should be a top-down exercise.

STAKEHOLDERS

At the time of the writing of this book, there is a self-help guru on TV named Dr. Phil. One of his primary precepts relates to taking ownership of the consequences of your actions. The idea of ownership — but in a somewhat different way than Dr. Phil means it — also has relevance in the arena of risk/uncertainty assessment.

Most risk/uncertainty models are constructed in an attempt to forecast something. For example, we might build a model that attempts to estimate the cost of environmental remediation in a given area (but this applies equally to marketing schemes, construction proposals, and any number of other endeavors). So, the fundamental question might be: How much will the cleanup of this area cost? The cost estimate depends upon many things, just one of them being stakeholders. Each stakeholder can attempt to claim ownership of the requirements.

Stakeholders come in various guises. Some are political in nature. This type of stakeholder often manifests itself as a regulatory agency or a governmental body. In the U.S., the Environmental Protection Agency (EPA) is an example. These stakeholders typically impose standards to which you, as the responsible entity, will be held. Different stakeholders can, even for the same geographic area, have different and sometimes diametrically opposed directives.

Another type of stakeholder can be more cultural in nature. Natives to the area, for example, can have a significant impact on your remediation plans. So too can environmental action groups and other organized entities. The needs and wishes of such groups can also run counter to one another. For example, an environmental action group might require that you return the landscape to its original condition including the removal of roads. Natives, however, might wish the roads to remain because such roads can form the basis for new commercial enterprises.

So, here you are, trying to estimate the cost of remediation (or construction, or whatever). Various governmental agencies might be issuing conflicting decrees regarding how you will approach the project while organized groups and native inhabitants press to have their viewpoints honored. This is a dilemma.

I can offer two pieces of advice to help remedy such a situation. First, do your homework early. Attempt to identify the various stakeholders and their issues. If practical, convene meetings at which representatives from each group can exchange views and, hopefully, come to a consensus. Have such meetings facilitated by an independent and disinterested party — *not* by you.

A second piece of advice is to build a risk model that is robust enough to deal with the diversity of opinions. For example, Stakeholder 1 might have you perform tasks A, B, and C in your remediation effort, while Stakeholder 2 would have you perform tasks A, C, D, and F. In your risk model, establish cost ranges for each of the tasks. Also, estimate the probability that a given stakeholder's view will prevail. The integration of the cost ranges and probabilities in a Monte Carlo model will afford you a probabilistic cost range for your effort.

If the actual work performed is a mixture of stakeholder demands, this estimate might be close to the actual costs. If, however, a single stakeholder prevails, this estimate will not match actual costs, but might still afford a reasonable pre-settlement estimate of costs and will make transparent the tasks and costs associated with each stakeholder's demands.

SELECTED READINGS

Atkins, S., *The Name of Your Game — Four Game Plans for Success at Home and Work*, Ellis & Stewart, Beverly Hills, CA, 1993.

Jasanoff, S., Bridging the two cultures of risk analysis, *Risk Analysis 2*, 123–129, 1993.

Lam, J., *Enterprise Risk Management: From Incentives to Controls*, John Wiley & Sons, Hoboken, NJ, 2003.

Nichols, N., Scientific management at merk: an interview with CFO Judy Lewent, *Risk Analysis 1*, 89–99, 1994.

Pasmore, W. A., *Creating Strategic Change — Designing the Flexible, High-Performing Organization*, John Wiley & Sons, New York, NY, 1994.

Wheatley, M. J., *Leadership and the New Science — Learning about Organization from an Orderly Universe*, Berrett-Koehler, San Francisco, CA, 1993.

4 The Perception from Within — Why This Can Be So Darn Hard to Implement

SIMPLE RESISTENCE TO CHANGE AND MOMENTUM OF OLD WAYS

It is a rare circumstance that one is trying to initiate a risk/uncertainty process in a virgin situation. That is, there typically are at least segments of the organization in which risk/uncertainty-related processes already are implemented, or the business or technical arenas into which the new risk/uncertainty processes need to be injected have long histories and existing processes with nearly insurmountable momentum. Simple resistance to change is not only human nature, but can be augmented by organizational considerations. Rather than weave the many organizational and technical arguments on this subject into a story — as is my wont — I determined that a direct and decidedly more curt approach to the subject would be more effective. Thus, the subheadings below.

IF THEY DO NOT HAVE A SYSTEM, FIND OUT WHY

If a given business unit does not employ some sort of risk/uncertainty process, it is important to discover why they do not. There can be many reasons why a business unit might not yet employ some sort of risk/uncertainty system. Lack of resources, recent reorganizations, and other impacts might be rational reasons why some type of system is not used. However, there can be more sinister rationales.

One reason that a risk-related system is not used might be that there is a staunch disbeliever somewhere in the chain of command. It is important to identify such individuals before you approach the business unit with a plan to implement a risk/uncertainty system. *Why* this person is recalcitrant is all-important. Many times, such an individual was in the past a victim of poorly implemented plans foisted upon them. Whatever the reason, it is critical to discover these people and their motivations so that any plan you design will accommodate their viewpoint. Such individuals need to be addressed head-on — with kindness and understanding. Ignoring these individuals and hoping the process can roll them over is a plan for disaster.

Another salient reason that a business might not utilize a risk/uncertainty approach is a dearth of risk expertise. Some businesses are very close to the pointy end of the spear. That is, they work every day on the actual money-making projects in the field and are oft times overwhelmed with practical and real-world problems and practices. For such business units, it is difficult for them to see how the addition of a risk-based approach to their jobs would not be an impractical nuisance. In addition, the issue of pride can be a barrier. Many times, personnel in these business units are aware that other corporate entities are, in fact, utilizing risk-related processes and that their business unit appears to be relatively late in entering the game. As mentioned above, a dearth of risk expertise within the business might be the reason for failure to implement. It can be injurious to the pride of such business-unit personnel to admit and face this. Approaching such a business with a plan to implement a risk process should adroitly accommodate such considerations. Direct praise for their business acumen and their willingness to consider change and accept help can go a long way toward acceptance.

WISELY SELECT YOUR ENTRY POINT

Sometimes within a business that does not yet embrace risk-related processes, there is some individual in the organization who has been a lone advocate of implementing such an approach to business. I have found that although it is tempting to initially enlist the enthusiasm and help of such individuals, it can be a mistake to do so.

Start at the head of the business unit. Make your arguments starting at the top. If you begin your campaign with the lower-ranking enthusiast, it can be perceived as subterfuge and as going around the chain of command. This can be a critical mistake.

Business unit leaders often are well aware of the advocates within their organization and will suggest that you enlist the help or counsel of these people. If it is suggested by the business-unit leader that such an alliance be formed, then it might be to your advantage — but it only *might* be. Before identifying yourself with that individual and incorporating him into your strategy, be sure of two things. First, be sure that his idea of a risk-related implementation aligns with yours. Sometimes such in-business advocates have very different and sometimes — relative to you — diametrically opposed viewpoints. Second, be sure that this person is well received by others in the organization. Being aligned with the local wing nut is not necessarily an advantage.

MAKE A STRONG BUSINESS CASE

Although this subject is addressed in other parts of this book, it is well worth repeatedly assailing the reader with this point. Make sure you understand the business that you are approaching. Making examples of how other businesses have been successful in implementing risk processes is fine, but it typically will not sell a given business-unit leader on the idea. Examples of success from other businesses, though not without merit, are often met with the attitude, "Yeah, that is great for them, but my business is different and that example does not apply here."

Make it relevant. They will see through you in a second if you appear to be attempting to snow them. Learn the basic precepts of their business but *do not* try to come across as an expert. Admit what you do not know, but know enough to be conversant and relevant. The case you make for embracing risk-related processes *has* to make good business sense in that particular business and arguments about how such processes helped other unrelated businesses will be mainly lost. Realize that implementing a risk/uncertainty-related system will mean real change for them and more work. They will not like that. A solid and relevant business case that clearly relates how their business will benefit is essential. If you cannot come up with one, then perhaps either you lack a fundamental understanding of their business, or, they truly might not be able to benefit from adoption of a risk process. Again, these points are addressed in more detail in other parts of this book so I will not dwell on them here, but this is a critical aspect that could not be ignored in this section.

IF THEY HAVE A SYSTEM, THEY ARE PROUD OF IT

It is often the case that your view of what constitutes a risk system and that of a given business unit will not align. For example, the business might employ a risk/uncertainty process that helps personnel select contractors for projects. This is fine, but your idea of a risk system might have as its goal the probabilistic estimate of project value expressed as NPV that has been impacted by all relevant risk considerations such as legal, commercial, political, technical, financial, environmental, and so on.

If approaching such a situation, first find out as much as is practical about their existing system. Do not advocate that any system to which you might attempt to convert them will supplant their existing process. Rather, design an approach, when possible, that incorporates their system in your proposed process. For example, in the situation delineated in the previous paragraph in which a risk-based contractor-selection process already exists, find a way to incorporate that system into, say, the cost-estimating element of the more far-reaching process you propose. They likely are proud of their system and also are likely to have a host of users who understand and utilize the existing process. Take the approach of folding them into the new process rather than telling them that they will have to abandon the old way for a new one.

The bottom line is, attempt to discover as much about their existing system as possible so that familiar inputs/outputs and processes can be preserved to the greatest extent possible. Also, investigate how output from their existing process can be used as-is or modified for use in the overall scheme. Being sensitive to the issue of pride and that of user-base familiarity can go a long way toward acceptance of a new implementation.

HAVE UPPER-LEVEL SUPPORT

This might seem obvious, but the reader might be surprised at the number of times that promoters of new ideas and techniques — such as implementation of a risk-related process — do not enjoy the complete support of their own chain of management. Before you set out to try to convince business units that they should embrace risk-based processes, be sure that this advocacy is supported all the way to the top of your own management chain.

You can be sure that in some cases when you leave a business-unit leader's office, he will contact your management to determine the significance that the corporation is putting on this effort. It is not so much that they are checking on your credibility, but if compliance requires expenditure of significant funds and effort, that business-unit leader is going to seek assurance from corporate management regarding the seriousness of commitment to the idea. So, when the phone rings in any of the offices of managers in your chain of command, be as sure as you can be that the message they will convey is one of unequivocal support for the effort. If you cannot align such support, you might want to rethink your attempt to influence others.

PROVIDE HELP

Associated with the point in the previous section, you should be aware that many business units will be lacking in some critical aspect of a successful implementation. Sometimes they will not have available funds. Sometimes they will lack expertise. Whatever the shortfall, you should be prepared to at least offer suggestions — if not actual help — regarding how resources can be obtained.

Many business units will see this as an unfunded mandate. It is difficult to argue otherwise. You and your management should be prepared to offer real assistance in the form of money, personnel, or other resources. Just one more reason to be sure that you enjoy from your own management support for this effort.

EXPECT INITIAL REBUFF

Be nice, diplomatic, patient, and persistent. Initial reactions from business units can be to reject the idea. A hard sell in an initial meeting can be to your detriment. If you make the big pitch in initial meetings and your overtures are met with resistance or rejection, it is difficult to come back in subsequent meetings with new and fresh arguments for acceptance.

Initial meetings should focus on fundamental concepts and overall benefits to the business. These meetings should be relatively short and to the point. Short initial meetings in which they are not beat to death with arguments give business unit personnel a chance to reflect on the proposal and to get used to the idea. If rejection is the result of an initial meeting, subsequent meetings can be used to bring out the big gun arguments in an attempt to change their minds. If the big gun attestations are used in initial encounters after which knee-jerk rejection reactions are likely, you will have no fresh considerations to convey in subsequent meetings.

LEARNING HOW THE ORGANIZATION REALLY WORKS

Studying organization charts, if they exist, will give very little clue as to how a corporation actually works. Regardless of the solid- and dotted-line relationships on such charts, most internal operations are based on one or more of three primary drivers:

1. Necessity. Two or more corporate entities need to cooperate to survive. This necessity might be dictated by physical reality, as in the relationship between engineering/design and construction departments.
2. Personal relationships. Often, people in different departments have histories and have over time developed trusting relationships. Departments in which these people now work tend to be those that develop close working relationships — and sometimes, such personal relationships create terrific new linkages between departments that had little to do with one another before (law and finance, for example).
3. Edict. A dictate might be issued that necessitates cooperation between seemingly unrelated entities. For example, corporate executives might have noted that several high-exposure projects have, in the past, failed to achieve aspirations due to legal entanglements related to perceived or real environmental infractions. Therefore, it has been strongly suggested that any project proposal contain a joint environmental and legal assessment of risks and uncertainties.

An intimate and comprehensive understanding of just how the corporate mosaic actually functions can be essential in launching a successful risk/uncertainty initiative. Regardless of the impetus for interaction between seemingly disparate corporate entities, one should be aware of the characteristics of each type (necessity/personal/edict) and be prepared to exploit the respective traits in any attempt to implement a risk/uncertainty process.

When necessity is the driver, it usually is more likely that some sort of risk/uncertainty system exists — no matter how informal it might be. Such a system likely takes account of risk-related parameters associated with the cooperating departments. This can be a distinct advantage.

For example, in the course of executing multiple projects, an engineering/design department and a construction department might have developed a close working relationship. Seeking out and understanding these natural relationships can bring great economy to risk-process design.

An efficient risk process is one that respects relationships and presents parameters in a logic stream that emulates that of the real-world process. Much of the trouble with separately designed and built risk processes — such as, in this example, one for engineering/design and another for construction — is that outputs from one might not be readily utilized in another. This problem can be alleviated by creating a risk process that seamlessly allows data from one discipline to flow into a risk model for a second discipline. Decision making with regard to timing and budgets is greatly enhanced when, for example, the cost and schedule risk process naturally integrates cooperative entities such as engineering/design and construction.

When personal relationships are the impetus for cooperation, the entities brought together sometimes are less of a natural fit than those joined by necessity. Efficiencies need to be recognized and exploited whenever possible; however, the model designer has to decide whether addressing two groups as a single entity makes sense.

Suppose a natural real-world process progression passes data from group A to group B to group C to group D. Suppose further that because of a strong personal relationship between the business-unit heads of groups B and D a close working relationship has developed. The risk-model designer has to decide whether or not it makes sense to exploit the efficiencies in such a relationship in light of the fact that such collaboration violates the natural flow of materials and data in the real world. In such a case, it is best to determine if any of the efficiencies of this personal-relationship-driven coalescence can be utilized in a model that emulates the business relationships between the groups. If such efficiencies can be identified and can be fit into the natural flow, fine. If these efficiencies cannot be practically accommodated in the risk-model real-world-process emulation, then it is best to forgo incorporation of the practices.

One last note on personal relationships in corporations. Never underestimate them. Regardless of the lines connecting boxes on organization charts, most corporate operations are mightily influenced by personal relationships. People tend to cooperate and do business with other individuals with whom they have developed a trusting relationship. When trust is lacking between individuals in groups that should be naturally cooperating, such collaboration can be resisted and strained. When trust exists between individuals in groups that would not normally have a close working relationship, those groups tend to find ways to collaborate in business. Sometimes this leads to new business-enhancing relationships. Sometimes such relationships border on bizarre. With regard to how a business should actually operate, the risk-model builder needs to be savvy enough to cull those relationships that are fundamentally detrimental to the design of an efficient risk-based process.

Issuance of an edict can be a primary driver for cooperation. In many instances, such edicts are born of previous pain and suffering within the corporation. As mentioned previously, a dictate might be issued that necessitates cooperation between seemingly unrelated entities such as, for example, the law and environmental departments. Executives might have noted that several high-exposure projects have, in the past, failed because of legal considerations related to perceived environmental infractions. To remedy this foible in future projects, management might have issued a directive that requires a joint legal and environmental assessment of risks and uncertainties.

This is the type of collaboration that can look great from management's viewpoint, but might be perceived as a nuisance (or worse) in one or both of the groups being coerced to cooperate. Many such forced relationships can devolve into check the box tasks. That is, the two entities that perceive little common ground might for each project, for example, hold a cursory meeting at which data is input to a standardized form. This process allows compliance with the edict, but without enthusiasm and recognition of real-world impact, such thinly veiled cooperative efforts can yield little actual benefit.

This is not to say that all relationships that are prompted by directives are of lesser value. That is not so. However, such relationships can be perceived as substantially aiding the decision-making process, but might in reality be of limited usefulness. It is the responsibility of the risk-process designer to determine whether such relationships exist and, if they do, whether incorporation of the seamless flow of data in the risk model between these entities is of real benefit. The decision-making process is not naturally nor necessarily enhanced by blindly embracing such seemingly cooperative situations.

PRACTITIONERS WHO HAVE BEEN DOING IT FOR YEARS ARE DIFFICULT TO CHANGE — NEWCOMERS ARE EASIER TO DEAL WITH

If you do not know how to swim and I am attempting to get you to execute the backstroke to my stringent specifications, after learning the technique, you likely are going to be able to execute the stoke to my satisfaction. However, if you are an experienced backstroker, it is going to be much more difficult for me to change the way you are already used to executing the stroke.

So it is with risk/uncertainty processes. Those to whom this is a new experience need only be convinced that it is good and necessary and to be tutored in the execution of the process. If, however, you have traditionally performed a version of risk/uncertainty assessment or management that now requires changing, the challenge is much greater.

As mentioned in a previous section, it is a relatively rare instance in which a risk process is being considered in a business that does not already in some manner consider risk and uncertainty. Sometimes, these already-entrenched processes pervade the business to such an extent as to form the foundation of their business and decision-making process. However, these already-employed techniques might be less sophisticated and comprehensive than those required to allow that business to become part of, for example, a multiple-business or corporate risk-based process.

This is a real problem. Personnel in such businesses might consider themselves to be quite up on risk practices — whether or not that is actually so. If luck is with the purveyor of the new risk/uncertainty process, the existing business-embedded risk-related techniques might be able to be incorporated more or less as is into the new more comprehensive scheme. More often than not, however, this is not the case.

There are three practical approaches to resolving this problem. The first has already been mentioned. Try to find a way to incorporate the already-existing risk-related process into the new scheme. It this fails, the second alternative is to attempt to convince the business that changing from their existing process to the new techniques not only will benefit the business directly, but will have positive ramifications for the corporation as a whole. These arguments have to be real and true — it will take business personnel only a second to see through any thinly veiled attempt to persuade them with spurious arguments.

Sometimes, neither incorporation of the existing process nor influencing business personnel's willingness to change is possible. In such instances a third and last-resort enticement has to be employed — that of a directive from upper management. As previously delineated, the purveyor of the new risk process has to be sure that he enjoys the support of his own management chain and that the corporation truly believes that adherence to the risk-based business model is essential. If this is so, then sometimes it is necessary to force compliance. This will not enhance anyone's popularity, and such tactics should be used only as a last resort.

NOT INVENTED HERE

Pride is always a consideration. Whether or not a segment of the organization has a pre-existing process that can be perceived to be related to the new risk/uncertainty process, one can encounter stiff organizational resistance to implementation because the new process is being imposed from outside. This is a fairly common situation.

First, the person proposing the new risk process should attempt to foresee when such resistance is likely. In cases where it is foreseen, it is best to approach the business entity with the overall plan. That is, begin communication by delineating the big picture view and how important the process is to the corporation. Then enhance the conversation by describing — and it has to be true — how critical is the incorporation of a risk-based process for this business.

Follow these discussions with conversations regarding how they see their business contributing to the effort. These discussions should be detailed enough so that the business feels as though — and it might be true — it is designing its own piece of the overall risk process. Of course, a skilled facilitator will guide such discussions so that the ultimate design will be one that comfortably fits into the overarching scheme but will seem to have been designed by the business itself. This, of course, takes time but it is one of the only practical means of overcoming the not invented here syndrome.

INSTANT ACCESS TO EXECUTIVES — IF YOU ARE FROM OUTSIDE, YOU ARE SMARTER/BETTER

If you work within the organization, very likely there is a pecking order and organizational protocol that must be recognized. That is, delivering your message to the top of your organization might, at the very least, be seen as going around more immediate management and resented. However, if you are from outside the company, you can have immediate access to nearly any corporation decision-making level desired.

In addition, in many corporations, there seems to be the pervasive concept that if you are a consultant or otherwise external to the company, then your message and wares are more credible and desirable. (Support issues and costs can have a lot to do with this.) This perception, of course, is not necessarily true, but certainly is a fact of life inside a corporation.

Why are these things so? Fundamentally, a corporation operates like a family. Persons of great renown typically get the least respect from those who know them best. It is a bit too much to suggest that familiarity breeds contempt, but it certainly is true that familiarity fosters a jaded attitude. Newcomers with an apparently fresh idea seem to be more readily embraced than those familiar souls who are promoting ideas of equal or greater validity. These reactions to the familiar and to the new are simply human nature.

A convoluted or perverse reward structure also can be an impediment to in-house-generated ideas gaining equal footing with those from outside the corporation. Many times, in-house initiatives have to be sold up through the management ranks. If the idea is deemed a departure from the business norm, subsequent upper-level managers might be reluctant to promote the idea upward. This reticence is due, in part, to a fear that their reputation could be damaged by supporting a hair-brained idea and, therefore, their reward will be diminished. Such are the realities of corporate life.

If there exists a level in the corporation from which, for example, the idea of a corporate-wide risk/uncertainty process requires support, it can be an exceedingly arduous task for someone from within the corporation to reach the desired corporate level with an unadulterated message (a message that has not been modified — sometimes beyond recognition — by translation at multiple corporate levels). Again, if the critical corporate level for support can be identified, the outside consultant, for example, has the luxury of attempting to arrange a meeting directly with the appropriate executives. This is a distinct advantage.

Unfortunately, it is the author's opinion that there are few avenues of safe recourse for the inside-the-corporation purveyor of an idea. One avenue that is fraught with detrimental ramifications is to simply go around immediate management to reach the desired level. This, of course, is not only bad form, but the message delivered is likely to be poorly received by the targeted level of management. This tack is not recommended.

Another ploy that is not optimal but is likely more effective is to have the idea promoted by an outside entity. This, of course, might mean sharing credit with (or relinquishing credit to) a consulting firm. However, if the corporate person truly has the best interest of the corporation in mind, it might make sense to have the idea marketed to the corporation from outside with the hope that if the idea is accepted, he will be a primary contributor to implementation within the corporation.

Yet a third approach to selling an idea within a corporation is to enlist the support of peers and other respected individuals. If a consortium, of sorts, representing a broad spectrum of respected individuals comes forward to support a new idea, individuals at various management levels will be much less reticent to lend their support and to promote the idea upward. In addition, a consortium of like-minded people has a much better chance of delivering their ideas directly to the critical level of management than does an individual.

TURF WARS

A corporation is not unlike the proverbial pie — there are only so many slices of useful size. From the outside looking in, a corporation can appear to be a cohesive and unified entity. One peek under the tent, however, usually reveals an organization that is rife with internal struggles for power, control, and longevity. Nearly everyone is vying for their piece of the finite resource pie.

This internal strife applies to processes as well as political and organizational entities. It behooves the purveyor of a new risk/uncertainty process to be aware of existing risk processes and the power they represent for those who control the processes. Once again, as illuminated in the previous section, anyone introducing a new and potentially competing risk system or one that is perceived to replace existing practices should carefully consider the situation.

Implementation, unless by edict from above, will only be successful if it appears to those who are having the new system foisted upon them that the new approach either encompasses their existing process, or will enhance their recognition (deserved or not) as an expert in the risk arena. There is no substitute for a sound political strategy that recognizes and, to the extent this is possible and practical, embraces the existing processes.

It can also be the case that there are no prominent existing risk practices in use in the business entities to be impacted by the proposed new risk/uncertainty process. Introduction of such a process in virgin territory can initiate power struggles for control of the new system.

Corporations are not democracies. In the end, control of and responsibility for the risk/uncertainty process installed in a given discipline likely will rest with one business entity (one department or group) that is led by a management team. If the purveyor of the new risk/uncertainty process has any influence regarding where responsibility will ultimately rest, he should attempt to identify the management team that is most likely to be open to inclusion of multiple groups in the decision-making process and that will be as open-minded as possible regarding funding, training, and other implementation issues. Such management teams often are not those who would espouse to have the greatest prowess in the area of risk/uncertainty. In fact, sometimes those groups who have experience with lording over a risk process in the past can be most internally focused and prone to exclusionary practices.

Once again, there is no substitute for recognizing how the corporation actually operates. Unfortunately, recognition of natural and invented sources of internal tension is a salient aspect of self-education in this area. Creation of a sound strategic plan prior to any attempt to implement is essential.

EXPERTS WHO ARE NOT

If it becomes fashionable to integrate risk/uncertainty into the fabric of a business, any purveyor of a credible process likely will not be alone in attempting to influence the company. Unfortunately, no license or degree is required to declare oneself a risk/uncertainty expert. It is likely that the company will be besieged by experts who are not and by those who are knowledgeable in the science, but who do not appreciate the internal organization into which they are attempting to inject risk/uncertainty processes.

Some experts exist outside the corporation. These folks tend to be those who garnered some expertise in a narrow aspect of the broad risk/uncertainty arena. For example, in a major corporation, an individual might have gained great insight into how marketing risk/uncertainty assessments are conducted. Upon leaving that corporation for a consulting career, the risk expert finds that the spectrum of businesses to which risk techniques can be applied is very broad and advertises his skills as being applicable in diverse aspects of business. Rarely is expertise gained in one business arena directly applicable to other areas of business.

For example, a person in the exploration arm of an energy company might gain significant expertise in the area of subsurface risk/uncertainty. The individual might have gained expertise in the estimation of porosity ranges, permeability, migration pathways, trap volumes, recovery rates, compartmentalization of reservoirs, and other aspects of prospect risk assessment. Expertise in such things is a significant accomplishment and is not to be underestimated. However, such expertise is of very limited value when attempting, for example, to analyze the complexities of a legal litigation or when trying to integrate political risk/uncertainty considerations in a probabilistic assessment of complex contractual issues. Such things require an entirely different and respectively distinct skill set.

It should be recognized that any risk/uncertainty expert that would attempt to approach, for example, the problem of establishing a probabilistic method for corporate portfolio management of potential investments should have experience in handling risk/uncertainty assessments in the technical, legal, contractual, political, cultural, and other arenas. Individuals who have a broad background of experience with regard to disciplines likely can be considered credible mentors for corporate implementation processes.

The argument presented in the preceding paragraphs applies equally to internal experts. Often, the purveyor of the new broad-based risk/uncertainty process, when approaching a particular department or business, will be told by management of that business: "You really need to work closely with Joe — he is our risk guru and can be of great help to you." Come to find out that Joe is a real wiz, for example, at assessing and managing risk related to division orders, but has little experience outside this narrow field. Management of that business might not know of any other risk experts so, to them, Joe is the man. Not that Joe is not a nice guy, and not that he does not exhibit particular knowledge in his field of endeavor; however, such expertise will have value in the overall plan in proportion to the role that, in this example, division orders plays in the wider execution of the business and in the still broader area of corporate portfolio project management. Recognizing and working with the "Joes" of the world is a critical part of a risk/uncertainty implementation plan that has any chance of succeeding.

MANAGEMENT THAT HAS LIMITED APPRECIATION AND THE DETERMINISTIC MINDSET

Speaking of fashionable, as I was in the previous section, it often is the case that upper management will find it trendy to request risk-weighted data for projects while still expecting they will get "the number." Ranges, truly risk-weighted values, chances of abject failure, and other risk/uncertainty-related data typically are not appreciated nor fully understood by those to whom these things are delivered. Recipients often are not prepared to utilize this information to its fullest extent — if at all. I cannot tell you the number of times I have had conversations with individual managers and management teams at various corporations who claimed to be all in favor of a risk/uncertainty process in their business but who, at the same time, fully expected that all of the action would take place below their station and that implementation of such a process would not alter the way they did business. No other notion could be farther from reality.

In a later chapter of this book, I address the issue of training. Not to jump the gun, but it bears mentioning here that training should *begin* with management teams. In my younger days, I believed that training the troops who would actually perform the risk assessments was the way to start the process. I was mistaken. After a few years of pursuing this policy, it became clear that many of the people who were trained in the process of risk assessment for their business had ceased to perform such analyses. The primary reason for terminating the practice was that once they had generated probabilistic output (cumulative frequency curves, sensitivity analyses, etc.) there was no one up the chain who appreciated the strange information and few who could actually incorporate such information in their decision-making process. So, without someone to whom the practitioners could deliver their probabilistic wares, what was the point? Not only that, but they were pressed to deliver the deterministic-type information that they had traditionally created and served up. This risk stuff, therefore, became a distraction, and worse, a black hole for time, money, and effort.

So, I thought better of this approach. I learned — admittedly, the hard way — that the place to start is with education of those to whom probabilistic information will be delivered for the purpose of business decision making. It is currently my practice to engage the executives (i.e., the decision makers) in at least one-day seminars in which, by example and otherwise, I extol the virtue of the risk/uncertainty approach in *their* business. This training has to be specific to their business and relevant to their particular decision-making process. Presenting generic examples or stories from different businesses will not be seen as relevant to their practices. This, of course, means lots of work for me. It is difficult, sometimes, to learn enough about a particular business to enable me to create real and relevant examples to which the audience will relate. However, this type of preparation is absolutely necessary.

Such prepared presentations must exemplify why a probabilistic approach actually would benefit their business and help them make better decisions. For example, most businesses traditionally will set deterministic targets related to production (how much they will make or how much they will sell) and costs. Just one of the examples I typically present is one in which a deterministic target value falls on a Monte Carlo-generated cumulative frequency curve. Such a curve shows them the probability of achieving the target value. Sometimes such values are very easily achieved — i.e., not aggressive enough — and sometimes such values turn out to be quite impossible to attain under normal circumstances. Viewing the cumulative frequency curve can relate such information.

In addition, such curves relate to the decision makers the range of possibilities that might result from a particular decision. Often, people are amazed at, for example, the tremendous downside to which they are exposing their business. I point out that considering only a deterministic value (a mean or a P50 or whatever) gives no hint regarding the potential exposure of the business to this decision. This type of illumination, if presented correctly, is always appreciated.

In addition, I attempt to convey some of the less direct advantages of the probabilistic approach. One such advantage is the roll up of individual analyses. For example, it is not atypical that a business will gather deterministic production or sales forecasts from individual departments and sum them up in an attempt to gain insight into what the overall business might achieve. Unless the deterministic values are means, they cannot be legally summed, but I typically try to avoid rubbing their noses in that one. I show them how using a Monte Carlo model to sum the individual department cumulative frequency ranges will result in an overall business range of results and that how pulling single values from such a summary curve is much more meaningful. This has the advantage of letting them have their cake and eat it too. From such an exercise, I point out that they are in a much better place to make decisions when probabilistic information is generated and analyzed, but that they can also then extract single values (means, etc.) from such probabilistic projections for use in their traditional business reporting practices. This, I find, goes over very well.

So, the bottom line is that I find that if I can instill in them a real sense of worth related to a probabilistic risk/uncertainty approach, relieve them of any fears they had of such practices, and demonstrate that such techniques can, in fact, be practically implemented in *their* business, reception of the use of such supposedly hair-brained schemes is greatly enhanced. Decision makers have to *want* to do this. Without someone in management requesting such information from practitioners down the chain, implementation of such a process has little chance of success.

COMING FROM THE RIGHT PLACE

Unfortunately, corporate politics are such that it does matter from where a message comes. For example, in a pharmaceutical company, processes promoted from, say, the research arm might not be readily adopted by those in a marketing business unit. The proponent of risk/uncertainty processes should be organizationally and politically positioned such that his message will be received favorably by the target audience.

As discussed in previous sections, the right message must be delivered and the realities of corporate politics and of the actual way a corporation operates need to be heeded. All this said, there is at least one more significant aspect of implementation that is critical: finding out the place from which the effort emanates.

Neutrality has its disadvantages and advantages. If the proponent of the new process appears to come from a neutral corner, that is, a noncompeting (for resources, power, control, or longevity) entity such as a corporate office, then the message has a better chance of being heeded. This can be for two reasons. One is that the corporation is the source of resources, so it is not in any business-unit leader's best interest to bite the hand that feeds. A second reason is that a corporately-derived message might have a more warm reception if it comes from a source that is not competing directly with the business unit for resources. In addition, there can be a fear on the part of the business unit that other businesses are being approached with the same message and that their business should not be seen as the odd man out or lose advantage to other businesses by late implementation of the proposed process. Also, some resistance can stem mainly from pride. Being told what to do by the corporation, for some psychological reason, does not carry with it the sting to the pride nor the stigma that comes with being set straight by a competing entity.

The old adage "Hi, I'm from the government and I'm here to help" exemplifies the attitude many business-unit leaders might harbor toward those who visit from the corporate office. Typically, corporate oversight and perceived meddling in business-unit affairs might not be appreciated. A distinct disadvantage of having a message delivered from a neutral corner such as from a corporate office is that individuals representing such offices can be viewed by business-unit personnel as not knowing what they are talking about — especially when it comes to the operations of that particular business. Corporate guys can be perceived as having been out of the game too long and their opinions and insights no longer relevant. The only counter to this is to be sure that it is not true. Be sure to do the appropriate amount of homework regarding knowledge of the business prior to approaching that business.

In any event, it has been my experience that there do exist peer groups and other cooperative bodies that share information between business entities. All is not cut-throat competition (as exemplified by the cooperation of departments based on personal relationships, delineated in a previous section of this chapter). However, the reward system in most corporations is set up to reward those who succeed — sometimes at the expense of others. Coming from a place that can only benefit from the success of a particular business — such as a corporate office — does alleviate somewhat the reticence to receive and consider new proposals.

So, the bottom line in this section is to consider the impact of the *source* of the message on the receptivity of a business unit. Try to see it from their perspective and position the message source such that it will be met with the greatest amount of enthusiasm in the individual businesses.

ESTABLISHING CREDIBILITY SO YOU ARE NOT JUST ANOTHER VOICE IN THE WILDERNESS

Corporate executives — especially in corporations that are in trouble — get advice from all directions. Although a credible risk/uncertainty message might truly help the company, it should be realized that to those executives, the proponent of such a message is but one more voice in the wilderness.

Some people who advise the corporation that it adopt, for example, a risk-based portfolio management system might enjoy a high-profile and highly respected position in the corporation. If this is the case, then advantages and leverage associated with that post should be exploited. However, from the perspective of the top of the house, most people with good but somewhat radical ideas typically emanate from the ranks.

So, lets assume that the idea of creating within the corporation a risk-weighted project evaluation scheme begins in, looking down from the top, the morass that lies below. How do you get your message heard?

Again, unless the proponent of the risk process enjoys some political in, the very best means of prompting executives to pay attention is by demonstrating, on an actual project with which they are familiar, that the proposed methodology actually exhibited a positive impact on that endeavor. On the part of the risk-process proponent, this takes time and effort up front. However, the impact on the corporation — if this methodology is adopted — will be huge and, therefore, the effort is warranted.

First, the risk-process proponent should identify a real-world project that is important to the corporation. Next, identify individuals within that project who will be primarily responsible for its outcome. With those individuals, or individuals reporting directly to them, attempt to arrange short and concise meetings in which, by demonstration, the benefits and costs of the proposed approach are clearly delineated. The demonstration should include examples that directly relate to the project in question. Generic examples or analogies from different businesses, while excellent testimonial material, are not likely to persuade those individuals to carry your message up the chain.

This approach is a lot of trouble. It obligates the risk-process proponent to learning enough about the targeted particular business to be able to produce an example that will pass the laugh test. In addition, the people who are being asked to take this proposal forward will have a plethora of questions that have to be addressed in terms and in a manner that they do not have to translate. On the part of the risk-process proponent, learning the language of that particular business (and believe me, each individual business has its own terminology and alphabet soup of acronyms) is essential. However, as mentioned several times in this book in various places where I address the issue of client interaction, the proponent should *not* try to come across as an expert in the area. This will be interpreted as having learned all there is about this business in a week or so. This is an insult to those who have spent years actually learning the business.

Creating a credible example can mean having to obtain information about the business that is not general knowledge. This can mean having to enlist, early on, a colleague in that business who can aid in the collection and processing of business-particular data. Identification of such individuals can be critical. Such people can not only help collect needed information, but can help translate the risk-process output into business-digestible text.

Presentation of the risk-process output should be comprised of a presenting team that includes individuals from the impacted business. Any presentation should clearly demonstrate the costs and time required to implement the new process and demonstrate in an absolutely crystal clear manner the benefits of such an approach. Many times, such as those in which risk-weighted portfolio-management processes are being proposed, the benefits are more fully realized at the corporate level and not at the level of the business to which you are presenting (i.e., "If you go through all this trouble to produce risk-weighted values for your projects, the people at the corporate level will be able to better compare the value of your project with that of others. Is that not great?"). Although true, this is hardly a selling point that will entice business-unit personnel to embrace your proposal.

Just some of the benefits that the particular business should glean from the proposed risk-based approach are:

- Clarity regarding just what question is being addressed. That is, just what is our business objective? A well-orchestrated risk process will, as a first step, make clear just what is the goal of the exercise. See the What Is the Question section later in this book.
- Clarity regarding just what information needs to be addressed to arrive at the desired answer. This not only brings enhanced understanding to the overall project team, but is beneficial in keeping them from pursuing information that will end up being extraneous.
- Capturing uncertainty regarding all pertinent parameters. This process can be exceedingly enlightening. Parameters that were previously taken for granted and difficult-to-talk-about variables should be openly addressed in the risk process. This allows business teams to consider best, worst, and most likely scenarios related to each parameter. The integration of these ranges in the final risk-model output generally has surprising results.
- Considering systemic interactions between parameters. When contemplating project influences outside the risk-model framework, such influences are typically considered in conceptual isolation. That is, each one is considered alone, sometimes by different groups of people, and it is very easy, using this approach, to overlook interactions between parameters. When considering those same parameters in the risk-process scheme, dependencies (i.e., correlations) and interactions between such variables is part of the process. Accounting for such systemic interactions can have a significant impact on the results of the analysis.
- Interaction of teams. This is a *very* important ramification of using the risk-based approach. For example, in order to capture information regarding the interaction of risk-model parameters, the facilitator of the process will insist that business-unit groups who typically fail to interact (law and commercial groups, for example) sit down in the same room at the same time and exchange views on the subject. This is of great benefit when attempting to build a risk-based emulation of the real world.
- Range of possibilities. Typically, when results of deterministic business analyses (i.e., spreadsheet-based analyses) are presented, *the number* is the answer. That is, a single value resulting from a single-scenario analysis is presented. To get a feel for sensitivity to changing parameters in such a model, many deterministic analyses might have to be executed. In executing such a scheme, there is, of course, no means of observing the impact of interaction of the various scenarios. When employing the risk-based approach, all scenarios can be incorporated and caused to interact. The result of this approach more closely emulates real-world situations.
- Clarity regarding the probability of achieving a goal. Related to the point above, when a deterministic result is generated (or a host of deterministic answers resulting from multiple-scenario runs of a model), the person

interpreting the results can have no feel for how likely it is that the business will realize the reported number(s). Results from a probabilistic analysis clearly demonstrate not only the likely range of outcomes, but the probability of achieving (up to or greater than) any selected point on the range. This is critical information for decision making.

These are but a few of the selling points that should be employed when attempting to convince a particular business that adoption of the risk-based approach will actually have direct benefits to the business (and not just to the corporation). The approach suggested in the preceding paragraphs is a lot to ask. It is time consuming. However, it is the only approach I know that, if skillfully executed, will certainly get the attention of the corporate decision makers and will distinguish the risk-process proponent's voice from all others in the wilderness.

TIME FOR PREPARATION OF THE ORGANIZATION — WHO BENEFITS?

Risk models sometimes are designed for and executed on single opportunities in a one-off mode. More commonly, risk assessments are performed with the goal of comparison in mind. That is, we generally assess not just one opportunity, but several with the aim of comparing these opportunities so that we might manage our portfolio of options. When this process of comparing and culling opportunities takes place in an organization, it necessarily begets winners and losers.

Most organizations need time to adapt to this mindset for several reasons. First is the generation of winners and losers (to be addressed in detail in a later section). Typically, prior to implementation of a risk model and the resulting comparison of opportunities (allow me here to call these opportunities projects), each project might have been prepared and carried out in relative ignorance of other seemingly unrelated projects. Each project, then, gained support and funding based on its individual merits. With the advent of risk assessment and portfolio management, everything changes. The new process necessarily ranks projects relative to one another on some common ground, for example, net present value (NPV) to the corporation. Under the new risk/portfolio management scheme, projects that were, in the past, considered to be unrelated are being compared to one another. Funding is being given to some and denied to others.

The risk assessment and portfolio management scheme might appear to be a great boon to the corporation. For example, those who sit in positions that make them responsible to stockholders might see this process and the resulting increase in efficiency and revenue as a great step forward. Not so from the perspective of those whose projects are being risk-assessed and portfolio-managed. From their perspective, this process takes valuable time, often brings to the fore uncomfortable feelings and situations, causes them to quantify and document things that were heretofore taken for granted or assumed, and results in projects not being funded. The project upon which they were working might be passed over for another project of which they may never have heard.

Under a more conventional project selection and funding scheme, most business units (groups) could be assured of some level of project funding in a given year. With the advent of the risk ranking/portfolio management process, funding is no longer likely to be more or less distributed over all projects. Some will be selected for funding and some will not. Even those project proponents whose projects are selected are likely to resent and/or resist the implementation of a risk process.

Prior to this new process, projects that were funded were not necessarily compared with other dissimilar opportunities. Implementation of the new risk-related process not only compares a project to a wider range of competing projects, but the new process forces project proponents to comply with the quantification of project parameters and other rigors of the imposed risk process. Therefore, even those groups or individuals whose projects are selected for funding feel as though Big Brother is watching and that they are no longer in control (if they ever were) of the situation. In addition, funding levels are likely to be similar to funding levels in the past, but those who get funding have to go through much more work than they had previously (for the same result).

So, from the corporate management point of view, this looks great. From the point of view of the people associated with the project, this looks like a process that is a lot of work and results in a significant chance of not being supported. It turns out that no amount of talking on your part (the instigator of the risk process) is going to convince those who have been hurt by the process, or those who could potentially be hurt, that this process is for the greater good and is going to be helpful to them in the long run. Time and continuous interaction with these people are the only remedies for this malady. Only when they are convinced (and this has to be true) that portfolio management of opportunities will result in a stronger company and that the risk assessment process really is a benefit to them, will they buy into the process and cease resisting the change. Corporate culture changes take time. Be sure to allow plenty of it if you are likely to head down the risk assessment road.

RAMIFICATIONS OF IMPLEMENTING A PROBABILISTIC RISK SYSTEM

Human nature is a primary influence in the implementation of a risk-based approach to determining project value. While there is no question that the proposed process is of great benefit to a business, there are at least several inherent business practices and aspects of human nature that run counter to smooth implementation of the process.

THE REWARD SYSTEM

Think about it. For what do you get rewarded? In most corporations, a person is rewarded for making things happen. Even if, down the road a few years, the project results in a net loss of capital, when said project is launched there is much fanfare and instigators of the effort are recognized and, in some manner, rewarded. Sound familiar?

Look, this risk-based thing is all about recognizing — both good and bad — risks and uncertainties associated with a project. It is true that there exist risks that can both benefit and detract from the perceived project value, but it is the nature of the world that there are, for example, more things out there that can cause a project to cost more than there are things that are going to save heaps of money. So what?

It is the job of the risk-process facilitator to identify all pertinent risks — both good and bad — and to assure that those risks have the appropriate impact on project value. If the premise in the previous paragraph is accepted (that there are more downside risks than upside risks, and/or the impact of downside risks typically is greater than those of upside risks), then that facilitator is, in effect, typically working at diminution of project value.

Most projects these days are marginal in value. That is, the probability of making a reasonable profit from a given project is not 100% or, most of the time, even near 100%. There can be significant uncertainty concerning whether or not a given project's economic outcome will meet some preset corporate hurdle. So, pointing out all of the risks and uncertainties associated with a project is, in effect, pointing out how marginal the project really is. If, as stated in the first paragraph of this section, we are rewarded for making things happen, then those individuals (and their processes) who at the beginning of a project are attempting to illustrate the potential weaknesses in traditional project profit projections are people who are not likely to be appreciated nor will they realize a reward under the typical reward system.

The reward system in most corporations needs to be revamped to appreciate accuracy rather than activity. Right now, if a person makes a project fly, they typically are rewarded in a short period of time. When that project, ultimately, does not produce the revenue string originally projected, the proponent of the project has moved on and is in most cases not directly associated with the shortfall. A system in which reward is delayed and tied to accuracy of initial forecasts — the projections upon which the project was sold — would better fit in the risk-based-process world. Not that this does not have its own set of problems.

For example, if I am to be rewarded on, say, accuracy of prediction of my annual spend, I can game the system to be perceived as always accurate. It is not much of a problem to determine what my actual costs might be, set my proclaimed cost somewhat higher than that, and at the end of the year when I fall short of the projected cost, simply find something upon which I can spend the shortfall so that I end up right at my projected amount. Guarding against this game with the institution of an oversight committee which can pass judgment on how reasonable are initial cost projections is a step that might be advised if a reward system based on accuracy is adopted. The point is, if traditional reward systems are kept in place, those systems will always clash with the implementation of a risk/uncertainty-based system.

CANNOT EASILY OVERLAY THE NEW PROCESS ON THE EXISTING BUSINESS MODEL

Everyone in a business inside a corporation is in competition for a piece of the pie. That is, there is only so much money and other resources to be divided up among the businesses. It does not behoove any business-unit leader to make his business or individual project appear any worse than necessary — especially relative to other businesses or projects with which they are competing for resources.

I have not run into many business-unit leaders at any corporations who would not appreciate knowing just what level of return a given project might yield. Realizing this level of return, however, necessitates the incorporation of impacts from all pertinent risks/uncertainties in the project-evaluation process. Given that businesses are in fact in competition for resources, the only way to assuage the reticence of business-unit personnel to incorporate all of the risk/uncertainty impacts is to assure them that all other businesses/projects with which they are in competition also are implementing such incorporation. This can be a hard sell if there has not been such an edict issued from the corporation directing businesses to adopt such practices, or, if there has not been put in place the aforementioned oversight committee the job of which is to ensure that all business-project proposals include the impact of all pertinent risks. If these measures, or something akin, have not been implemented, then it is unlikely that business-unit personnel will be compelled to adopt a holistic and comprehensive risk-based process for estimating project value.

Another reason that adoption of a risk-based system can be difficult is that it can fly in the face of traditional and entrenched business practices. For example, let's return to the admittedly oversimplified example of the portfolio of 10 dimes lying on a table. Each dime represents a project to be executed. Execution of the project is the flipping of the dime. If the project does not experience abject failure (i.e., failure to be executed), then the coin is flipped. Heads might yield 11 cents and tails 9 cents. Given the 50% probability of each result and the fact that the probability-weighted heads- and tails-result will be summed (5.5 + 4.5), the EVS cost is 10. Independent of this, there is a 50% probability that the project might not be executed at all, or, that it will be executed such that it does not impact results in the timeframe being considered. This yields a project that, except for the cost of doing business, costs nothing.

In the traditional budgeting process, each of the business-unit leaders for each of the projects would want to have their 10 cents of budget apportioned to them. Then, as time went on — on the average considering the coin-flip example — half of the projects would fail to come to fruition. If there is some sort of corporate budget give-back scheme, then the corporation will recover the money originally apportioned to the now failed project, but would have lost the value of investing that money over the time it took for failure of the project to which the budget was originally allocated. This amounts to an annual leaving money on the table exercise.

Now, consider the implementation of a probabilistically based budgeting system. At the beginning of a budget cycle, the business unit leader considers the portfolio of ten potential projects. Each one could cost 10 cents, but the business-unit leader also knows that each project, in the flipping-the-dime example, has a 50% chance of failing to be executed (and, therefore, costing us nothing but the cost of doing business). So, the risk-weighted cost of each project, considering the 10 cent budget and the 50% chance of abject failure, is 5 cents. This yields a risk-weighted portfolio cost of 50 cents.

The business-unit leader turns to the corporation and has it cut him a check for 50 cents. Now that leader turns around, and what does he see? He is facing 10 project leaders each of whom expects to be allocated 10 cents for his project. The business unit leader does not have 100 cents — only 50 cents. Now what?

Under a risk-weighted budgeting scheme, for example, a new business-unit process would need to be implemented. No longer could each project leader expect to be given his entire projected budget at the beginning of the budget cycle. Each project leader would have to be assured that as he began to spend money, the business unit would cover the checks. This puts the project leader in the uncomfortable position of not actually having the budget money in the bank with which to negotiate contracts, buy supplies, etc. This process can work, but it should be kept in mind that it absolutely flies in the face of most accepted and traditional budgeting schemes.

Now, I realize that budgeting is but one part of the overall business process and that the dime example I have chosen to employ is overly simplistic, but this diatribe does illustrate the fact that implementation of risk-based processes will have some unintended and, certainly, unexpected consequences when an attempt is made to incorporate such processes in the traditional business scheme.

Two Sets of Books

Considering the corporate debacles and collapses of recent years, auditing of corporate processes and projections has taken on great importance. Adoption of a risk/uncertainty-based practice can have implications with regard to the audit process.

To return one more time to the portfolio-of-dimes example, let's consider that each of the 10 dimes again represents a project to be executed — execution in this case being the flipping of the dime. As described previously, the EVS is 10 resulting from execution of the project. Independently, there exists a 50% probability that the project will not be executed or will be executed too late, and the consequence is 0 cents. Again, as described previously, the EVP, therefore, is 5 cents.

Now, we might employ an engineer, for example, who is responsible for building the vessel that will have to hold the winnings of a successful execution of the project. So, that engineer would have to design and build a vessel large enough to accommodate 10 pennies because if we do execute successfully, we will have to be ready to deal with the results from success. The 10 penny value is the expected value of success (EVS) — that is, what we expect to realize from the project if we are successfull in executing the project.

However, this is not the only resultant value with which we need be concerned. The corporation might want to know just how many pennies they should roll up in the overall projection — and in the announcement to Wall Street — of what the corporation is likely to produce in the coming year. Considering that each project has a 50% chance of failing to be executed and, therefore, contributing nothing to the corporate coffers, the business-unit leader would not announce that his portfolio of dimes would be worth 100 cents to the corporation. Rather, the contribution of each project would be 5 cents and the overall portfolio contribution to the corporation roll up would be 50 cents. This risk-weighted projection of value is the expected value for the portfolio (EVP).

Well, now we have a project (dime) that has dual value. For planning purposes within the business unit, personnel have to work with the EVS. For example, people concerned with logistics have to sign contracts for train cars or ships or some other conveyance to move product from the production site to the customers. If the project is successful, it will produce the EVS value (in the dime example, 10 cents). We have to plan on and build for the amount of product we expect to realize if the project is successfully executed. However, the number the business would send up to the corporation for the not-yet-executed project has to be the success value (EVS) modified by the probability that the project might not be executed. This is the EVP. Now, to some auditors, it might appear that a business is keeping two sets of books because the business actually has two values for the same project. Auditing agencies will need to became comfortable with such projections of value when a fully risk-weighted process is employed.

FOR THIS TO BE EFFECTIVE, EVERY PART OF THE COMPANY MUST PARTICIPATE

A successful implementation of a probabilistic assessment/management process in any part of a company depends on the willful participation of all critical parties. The exclusion of one such salient participant can derail the effort. Universal participation is required for several reasons.

First, any assessment of a project's potential is akin to the proverbial chain — it is only as strong as its weakest link. For example, decision makers might be assessing the monetary benefit of a potential acquisition. The company's commercial, finance, logistics, marketing, and other departments have taken a critical look at the opportunity. Data from these corporate entities have been coalesced and integrated in our probabilistic model — the results of which indicate favorable, but marginal, potential. However, the law department is not convinced that the practice of quantitatively interrogating an opportunity is a good fit for their business and declined to participate. Their textual assessment, while part of the overall assessment folio, is not integrated into the calculation of estimates of value such as the expected value of success and the expected value for the portfolio.

Now, critics of the probabilistic assessment process will point out that it is not rocket science for corporate executives to take account of the legal verbiage when considering the opportunity. While this is true, this argument can be shortsighted. First, it is likely that decision makers do have the mental ability to integrate qualitative and quantitative information for a single project. However, if more than one department (such as legal) declines to participate, the mental gymnastics required to imagine the permutations and correlations between disparate information types become arduous, if possible at all. In addition, decisions are rarely concerned with whether or not a company will, in a vacuum, consider a single project. Typically, a project is vying for funding with competing opportunities. When multiple opportunities each require the mental integration of multiple value-impacting components, the process of portfolio management suffers.

Participation by all critical parties in the probabilistic assessment/management process is essential at yet another level. Consider the real-world situation of a project proponent. This person typically gets rewarded for making the project fly. This person knows that their project is in competition for funding with other opportunities. If the proponent of a project knows or suspects that proponents of competing opportunities are not participating in a probabilistic process that brings to bear on potential value all of the attendant risks and uncertainties, how in the world could the proponent in question be convinced to embrace such a value-damping process for their proposal? The answer is, they cannot be easily convinced to fall on the sword when nobody else seems to be interested in doing likewise.

As delineated in previous sections, if the probabilistic assessment/management process is not endorsed and expected by management, then it is hell on wheels to convince those who are in competition with one another — on multiple levels — to subscribe to a holistic probabilistic practice. For all of the reasons and more outlined in this section, it is essential to the successful implementation of a holistic probabilistic process that all critical parties involved be convinced that wholehearted participation is of utmost importance. Proponents of the probabilistic approach will find that the opting out of critical parties is much like the proverbial tiny hole in the dam. If people are not truly convinced of the merit of a probabilistic approach and are participating because of some sort of coercion, then the tiny hole of nonpartic-ipation can quickly become a flood of abandonment. As indicated in previous sections of this chapter, participants need to be truly convinced of the benefits of such an approach and need to feel sanctioned by those who mete out rewards that their participation is expected and appreciated.

NOT JUST ONE TIME — THIS MUST BE DONE OVER AND OVER, LIKE A DIET

At least parts of many corporations have a check the box mentality. That is, a given company section might tacitly subscribe to a corporate policy, but their heart is not in it. This lack of conviction generally leads to a check the box mentality and process.

For example, the head of the marketing department might not be convinced that a quantitative and probabilistic assessment of marketing trends, demographics, and so on really is warranted. However, compliance with a corporate edict that dictates such a risk assessment is required. So, as a final step before submitting the marketing assessment of an investment opportunity (and, if there is anything the reader should learn from this book, it is that the risk/uncertainty assessment should come early in the evaluation process), the manager of the department might gather together the same-as-always cadre of department personnel to take a half hour to enact whatever minimal process that will fulfill the probabilistic-assessment obligation. This gets their box checked. Now, the obligatory assessment is behind them and likely not to be revisited. This type of behavior can be commonplace, and is damaging.

Things change. As a project progresses, project personnel typically learn more about the unknowns and become more certain concerning parameters about which they originally harbored great uncertainty. In addition, some risks identified in initial meetings will, with time and effort, be mitigated. The impact of such risks and uncertainties on project value should be reduced as changes in their status occur. This, of course, demands at least periodic revisiting of the probabilistic assessment process so that new insights and data can be introduced.

This revisiting process flies in the face of the check-the-box mentality. Nonbelievers in the probabilistic approach need to be convinced that revisiting the assessment process several times during the project's pre- and post-sanction journey might actually be beneficial to the project and the perception of that project by decision makers. Like a diet, one cannot simply refrain from eating too much on one day — the diminution of caloric intake must be repeated on a regular basis.

The text above relates to the evaluation process for a single investment opportunity. However, the repeated application of a probabilistic assessment and management process also applies to the management of portfolios.

If corporate decision makers have adhered to the precepts proposed in this book regarding the calculation of expected value of success (EVS) and expected value for the portfolio (EVP) quantities for each opportunity in the portfolio, they will be able to use the relative magnitudes of the EVPs to help rank portfolio components. However, if project teams regularly update the EVP value for their projects, then it should be obvious that new portfolio rankings might result from re-evaluation of the updated information. So, this is like a diet at the portfolio level too — one cannot just do it once.

SELECTED READINGS

Berger, L. A. and Sikora, M. J., *The Change Management Handbook: A Road Map to Corporate Transformation*, McGraw Hill, New York, NY, 1993.

Davidson, J. P., *The Complete Idiot's Guide to Change Management*, Alpha Books, New York, NY, 2001.

Hiatt, J. M. and Creasey, T. J., *Change Management*, Prosci Research, Loveland, CO, 2003.

Kotter, J. P., *Leading Change*, Harvard Business School Press, Boston, MA, 1996.

Mourier, P. and Smith, M. R., *Conquering Organizational Change: How to Succeed Where Most Companies Fail*, Project Management Institute, Newtown Square, PA, 2001.

5 Consistency — The Key to a Risk Process

BIGGEST BANG FROM ASSESSING MULTIPLE OPPORTUNITIES

It is true that there is a benefit in assessing single and unique projects or opportunities with unique risk models. However, the maximum return on invested time and effort is only realized when multiple opportunities or projects are consistently assessed for the purpose of comparing, ranking, and portfolio-managing the assessed entities.

Within any organization, it is not strange to find that n number of projects of similar type are evaluated by n number of unique risk assessment processes and models. Typically, the assessment models are huge spreadsheets that have been handed down from generation to generation. It is not surprising to discover that the people using such spreadsheets did not write the spreadsheets and have only a cursory familiarity with the internal workings of the model. The quintessential user of such a model can feed it the information required as input and can capture and (sometimes) use and interpret the output, but cannot in detail delineate the process and calculations that link inputs to outputs.

Even within relatively small and focused business entities, it is not unusual to find that people whose offices are juxtaposed and whose tasks are similar are employing dissimilar methodologies to projects or opportunities of similar ilk. For example, within a group charged with the responsibility of evaluating the economics of plant (factory) construction, it is not atypical to discover Sally in Room 101 using a spreadsheet that calculates NPV from a 20 year time series of cash flows while John in Room 102 is employing a different NPV-generating spreadsheet that considers 25 years in its calculations. In fact, it is typical to discover that the spreadsheets, although evaluating similar factory-construction projects, do not require as inputs exactly the same information and link input values with output values utilizing different sets of equations. The unsuspecting manager of Sally and John believes that, when he receives the NPV values from the two subordinates, a valid comparison of the two projects can be made. The inconsistency of evaluation practices precludes any such comparison.

A unique assessment model applied to but a single project or opportunity makes good business sense when the project or opportunity to which the model is applied truly is one of a kind. For example, a corporation may be considering a stock split. A one-off risk model might be generated to assess the consequences of such a move. Such business maneuvers are not frequent events. The risk model constructed to assess the effect of the split, regardless of its comprehensive nature, may not be suitable for application to other business problems and will be used only once.

TYPE CONSENSUS MODELS

If it has been decided that a consensus model (or models) should be built (see Chapter 7), the logical next step is to decide just how many consensus models are required. For example, management of a construction company might decide that they could bid more competitively for plant-construction jobs and have a better chance at making a profit on jobs awarded if they could better estimate the range of construction costs and associated probabilities. Because their company, at any given time, is being considered for multiple projects, they foresee great benefit in consistently assessing the costs and manpower demands of the various projects. In addition, they would like to better predict the probability of being awarded a given project (they could be killed by success if they were to be awarded all projects upon which they have bid).

Management of the company might convene a meeting with department heads to discuss the plausibility of building such a model. In that meeting, management might discover that meeting participants believe that many such models might need to be constructed. It could have been the vision of management to build a single comprehensive plant-construction model, only to discover that the multitude of plant types may preclude this.

Department heads who work on chemical plants might voice the opinion that chemical plants are sufficiently distinct from other plant types to warrant a separate and unique chemical plant construction model. Waste treatment plant construction department leaders feel similarly about their project types, as do the department leaders who oversee power plant construction projects.

The problem, however, is even worse than this. People reporting to the head of chemical plant construction submit that separate models will be needed to evaluate dry chemical plants and liquid chemical plants. Similarly, those reporting to the head of power plant construction contend that the process of building a natural-gas-fired power plant is completely different from that of hydroelectric power plant construction. They argue that separate models are required. Waste treatment plant proponents intimate that building a plant that treats chemical waste is very different from constructing a treatment facility for solid waste.

This dilemma is not unique to the construction industry. An environmental firm might decide that they could better estimate cleanup and remediation costs, and compare and rank potential projects, if they consistently assessed their opportunities. The boss would like the ability to literally lay down on a table the results from individual assessments so that she could rank them relative to one another. However, when she proposes the idea of a comprehensive and consistent fate/transport-type risk model to the staff, the response of the staff is disheartening.

Those staff members responsible for assessing airborne environmental hazards contend that their charge could not possibly be assessed by the same model that is used to evaluate subsurface flow of fluids. In addition, the airborne assessors believe that they will need four separate models: one for healthy adults, one for healthy children, one for adults with respiratory problems, and one for children with reduced immunity. Likewise, the fluid transport assessment team members believe that they will need at least two models. They contend that separate models for toxic and nontoxic hazards need to be constructed.

When faced with such dilemmas, whether building plants, evaluating environmental hazards, or attempting to risk-assess entities in any other enterprise, it is essential that someone be empowered to determine just what will be the purpose of the risk assessment model. In the section What Is the Question? — Most of the Time and Effort in Chapter 7, I address in detail the process of how to discover what the real question is, so those details will not be reiterated here. However, it bears mentioning here that the managers of both the aforementioned construction and environmental firms need to decide just what it is they really want to know.

Subordinates at both firms argue that what they do is unique and warrants special consideration. This is typical. Management, however, contends that the firms need to do a better job of ranking and prioritizing their projects across the board. Both camps have a valid point.

It is a cogent contention that different types of plants and various sorts of environmental problems warrant specific and distinct risk models only if the objective is to assess in detail each of these seemingly unique entities. However, if the aim of the risk assessment is, as management would like, to create a capability to manage the diverse portfolio opportunities, then a general model that can accomplish this task must be assembled. Well, just how do you go about that?

In Chapter 7 and Chapter 8, I delineate the process by which the critical elements of a risk model are defined and assembled. The astute manager will first assemble representatives from groups that purport to require specific risk models to determine just how many truly unique entities actually exist. An extensive knowledge of the business, insight into personalities, and a grip on available resources and time should all be utilized in deciding just how many truly separate situations exist.

To evaluate each unique opportunity type, a distinct type model may need to be built. The savvy manager will enlist the aid of a facilitator skilled in guiding groups through the process of building a risk model. This process begins with deciding just what is the question (see Chapter 7) and includes the step of constructing a contributing factor diagram (see Chapter 8). It should be emphasized here that unique risk models should not be created for each project, but for each type of project. All projects of similar ilk can then be consistently evaluated.

When each group has successfully completed at least a preliminary contributing factor plot, the manager whose wish it is to compare, rank, and portfolio-manage the menagerie of distinct projects should collect and compare the individual contributing factor diagrams. The purpose of the comparison should be to identify common factors in the diagrams. Because it generally is the details of different project types that make them unique, the individual diagrams typically contain a multitude of components (even if they are labeled by different names on the individual diagrams) that are common to all diagrams.

For example, at the environmental firm, the facilitator may have led the groups to the consensus that they actually required four type models. They decide they need one to evaluate the effect of airborne contaminants on healthy humans, another to assess ramifications in humans with some susceptibility to airborne maladies, another to assess subsurface flow and migration associated with toxic substances, and a fourth model to consider remediation associated with nontoxic subsurface sub-stances. Upon inspection of the four contributing factor diagrams, the manager should notice a host of common elements. She might note that each diagram contains a variable that collects the number of internal man hours the project will demand. Another common parameter is the cost of external firms. Three more shared elements are the probability of governmental intervention and delay, the length of such a delay, and how much such a delay might cost in terms of dollars per day. Two additional common properties are the total number of days allotted for the project and the number of dollars per day the environmental firm might charge. Still another common element is the beginning date of the project. A list of these and other shared elements should be prepared.

Now, the manager herself has to go through the just what is the question exercise. If the manager would like to compare, rank, and portfolio-manage the projects based upon timing and potential profit associated with each project, then it should be determined by inspection whether the assembled list of common factors afford the capability of comparing projects based on the selected criteria. If it is determined that comparison is not possible, the manager should either ask the individual groups to collect the necessary additional information in each of their models or decide that she can live with a comparison of models based on the available list of common elements. When an acceptable set of common parameters has been assembled, the manager should build a contributing-factor diagram to serve as the basis for the project-comparison model.

Building type models is a salient first step toward consistent risk assessment and meaningful portfolio management; however, it alone does not assure success in risk-process implementation. Several insidious bugaboos need to be dealt with, not the least of which is what I call the counting the chickens problem (see Chapter 8).

I illustrate the counting the chickens problem by envisioning five houses built in a circle. The houses, therefore, share a circular backyard. In the yard are a dozen chickens. We knock on the front door of a house and inquire as to the number of chickens they have. The occupants glance out the back window and report that they have a dozen chickens. Repeating this process at each house leads us to believe that 60 chickens exist.

So what does the quaint chicken story have to do with real-world risk assess-ment? Consider the construction firm that has distributed a type construction risk model to each of its divisions. Each division is about to propose a year-long con-struction project. Each project will require many tons of concrete to be delivered over the year. The amount of concrete required by each project will stretch the sole concrete-supply company to the limit. However, each division, in ignorance of other divisions, populates the risk model with values that, taken alone, look reasonable. Unless some organizational mechanism is employed to ensure that the corporation is not expecting to have n times (n = the number of construction divisions) the amount of concrete the sole provider can deliver, each construction plan, and likely the corporation itself, is in jeopardy.

Just one more (of many) forms the counting the chickens problem takes is that of customer counting. It is not uncommon, in large corporations, to find that far-flung offices around the world that produce similar or identical products have targeted the same populations as primary customers for their product. Again, only organizationally driven coordination efforts can hope to avoid this counting the chickens problem.

COMMON MEASURE

The ability to rank, compare, and portfolio-manage risk assessed opportunities is predicated on the assumption that the assessed opportunities can, in fact, be compared. Ranking of opportunities requires that all opportunities be compared on the basis of one or more common measures. Typically, these measures are financial, such as net present value, internal rate of return, and others. However, measures that are not financial also are common. Parameters such as probability of success (technical success, economic success, etc.), product production, political advantage, and others can be used to compare opportunities.

If the goal of the risk assessment program is to assess individual opportunities or projects for the purpose of ranking, prioritizing, and portfolio-managing the assessed entities, then consistency in assessment is paramount. Consistency in the comparison of opportunities or projects of similar ilk is achieved by assuring that several things are true.

First, all opportunities or projects of a given type need to be compared using a common measure or set of measurements. For example, risk models that culminate in a financial analysis often generate risk-weighted net-present-value (NPV) plots, risk-weighted internal-rate-of-return (IRR) plots, or plots of analogous financial parameters. Most risk models result in the calculation of distributions for several such values which are then algorithmically combined to produce a single representative measure. This resulting parameter can subsequently be used in the portfolio management process.

In addition, each model must consistently generate the common measure. For example, two risk models may culminate in a NPV measure; however, this does not mean that the individual NPV values serve as the basis for a valid comparison. For example, one of the most common inconsistencies in calculating NPV is the number of years over which projects and their NPVs are considered. It is not atypical for a given risk model to calculate NPV from a 10 year projection of cash flows while another NPV-generating model considers cash flow over 20 years. Inconsistencies like this and others can make folly of the use of a seemingly common measure.

Although a noble aspiration, attaining a common measure from divergent risk models is not always practical. For example, there may be valid reasons why one risk model should generate an IRR measure while it makes more sense for another model to culminate in an NPV calculation. In still other situations, it may make business sense for one risk model to result in a measure such as tons of product produced while another model generates a financial measure such as NPV.

In situations such as these, if comparison and portfolio management of these opportunities is paramount, you should attempt to create normalizing routines that will convert one or more of the risk-model output parameters to a common measure. Examples might be algorithms that convert IRR values to NPV (with, perhaps, the collection of some additional information) or a routine that converts tons of product produced to cash and, eventually, NPV. The employ of such normalizing or conversion routines will be necessary if portfolio management of multiple and diverse opportunities is the goal.

RISK ASSESSMENT VISION

The preceding two sections of this chapter should lead to a vision and process for an organization that will afford it the ability to consistently assess and subsequently portfolio-manage its set of opportunities. Defining a vision of how you think this whole risk effort should work and where it should go is absolutely essential. It is important, when called upon to implement a risk process, to at least appear, after a reasonable time, to have a sense of what route the effort will take and where the work should terminate. Communication of the vision bolsters the leadership position, lends credibility to the plan (and buy-in), and lends focus to the entire undertaking. These are critical elements of successful implementation.

Once again, as the title of this book portends, the vision must be seen as practical. The initial steps of a vision should include steps to which people can relate and that they believe can be achieved. Setting too lofty a goal as an initial step can be discouraging rather than motivating. Too paltry a goal also can damage the credibility of the plan and can be interpreted by implementers as patronizing. Initial goals of the appropriate caliber will motivate people to achieve them and will impart a true sense of accomplishment when the initial goals have been realized.

Communication of this vision is essential. Often, a graphical representation of the plan is helpful, such as the diagram in Figure 5.1. The vision should be as

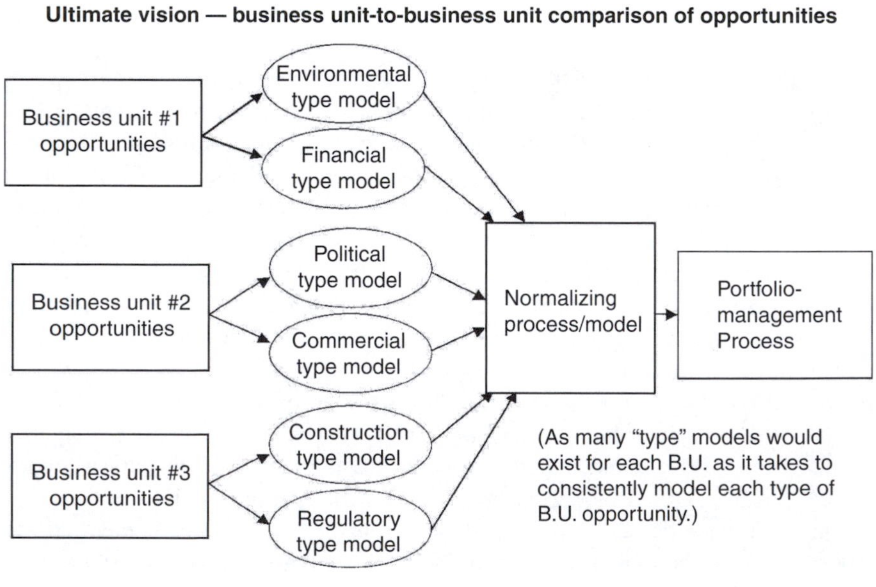

Ultimate vision — business unit-to-business unit comparison of opportunities

FIGURE 5.1 Risk assessment vision.

nonthreatening as possible and should include enough detail so that those to whom it is presented can see where they play a role.

Implementation of such plans is bound to cause some people or groups to eventually change the way they presently do things. The plan should be presented in such a way that the processes by which they currently do things fit into the initial proposal. People and groups who will have to change their ways should be presented with a rational plan spanning a reasonable amount of time so that they can feel as comfortable as possible with the transition.

In Figure 5.1, the ultimate vision is to generate a risk-based system and process by which all opportunities and liabilities of any type within an organization can be assessed, normalized, and portfolio-managed. Greater efficiency and maximum financial return are the aims. The first step toward this terminal goal is to establish, within subsets of an organization, the likely number of tasks or processes that are repeatedly executed. For example, within a law firm, attorneys may regularly decide which cases can be assigned to outside counsel and which must be handled in-house. The same attorneys may repeatedly decide, on a case-by-case basis, which legal actions to settle (and for how much) and which to take to court. Similarly, at a stock brokerage firm, analysts might daily decide which stocks to pursue. They may also regularly decide the proportion of funds that should be directed toward different types (bonds, money market accounts, annuities, etc.) of investments. The applications are, literally, endless.

When at least a few such tasks have been identified, risk facilitators should sit down with the various groups to work through the steps to build a consensus risk model and process (a process explained in detail in subsequent chapters). Care should be taken to ensure that multiple models do not overlap in their scope or intent. These models are depicted as ovals at the far left of Figure 5.1.

It is far more advantageous to do a few things right than to exhibit mediocrity at many things. When attempting to implement the type models, select a small number of areas that you deem to have the best chance of success. Proliferation of the vision and process throughout the organization, if done right, will follow. The very best way to encourage others to follow the process is to make heroes of the individuals or groups that have embraced the plan. Leading by example is the truest and most effective means of encouragement.

In spite of your obvious good looks, charm, wit, and technical prowess, this vision will not go forward without support from those in positions of authority. Buy-in to the plan from organizational leaders is essential. Communication of the vision to such groups or individuals is of paramount importance and should precede communication of the vision to the troops.

The second part of the ultimate vision outlined in Figure 5.1 is a normalizing routine. If portfolio management of opportunities is the ultimate goal, then this process likely is necessary.

Various type models within an organization will assess different entities. For example, a plant-production model might terminate with a distribution representing product produced per day. A legal model could calculate the amount for which the firm should settle a case. Still another model might assess the net income of a proposed new manufacturing facility.

What one would ultimately like to do is, figuratively speaking, to lay the output risk plots for each of these opportunities on a table and be able to identify the project that will return the most money to the corporation, the second best opportunity, and so on. Well, how do you do this when output on one plot is expressed in tons per day, another is shown as dollars lost to a settlement, and still another is expressed as dollars of net income per year? The answer is, you do not.

When generating the risk models, every effort should be made to have the models terminate in a common set of measures and units. Typical targets are financial measures such as net present value (NPV), internal rate of return (IRR), and others. It is not, however, always practical or possible to bring groups or individuals around to the realization that it is important that their model share some common characteristics with what seem to them to be completely unrelated risk models.

Portfolio management of diverse opportunities requires a common set of measures upon which those opportunities can be compared and ranked. The normalizing step shown in Figure 5.1 represents algorithmic annealing of various measures. This process can also require collection of additional information from various groups. The result of the process, however, should be a common set of measures (NPV, IRR, or whatever is important to you) that will allow comparison, ranking, and portfolio management of the collection of opportunities.

SOMETIMES BETTER TO BE CONSISTENT THAN TO BE RIGHT

Certainly it is of great importance to be accurate and precise when your risk model attempts to predict toxic exposure to humans, estimates project costs, or forecasts any number of other parameters that are sensitive to the magnitude of the output-variable coefficients. There is no question that the aim of such risk models should be to predict as accurately as possible both the magnitude of the calculated variable (x-axis coefficients on a cumulative frequency plot) and the associated probabilities (y-axis coefficients on a cumulative frequency plot). However, there is another class of risk-related problems in which striving for accuracy and precision is of lesser importance and, in fact, may be a completely unmanageable pursuit.

Let's first consider the simple problem of ranking a group of adults by height. My boss has been through the just what is the question process. Because his job is to arrange people on a stage for a choral presentation (tallest people in the center of the back row, etc.), he has decided that he is interested only in rank-ordering the people with respect to their height. He has delegated to me the task of assigning the members of the choir a place on stage according to their height. I realize immediately that I have two problems. One is that I will not see all of the members of the choir together again until the night of the dress rehearsal; however, they will be stopping by individually throughout the day to pick up their costumes. The other problem is that there is no accurate measuring device at the theater, and I have no practical means of obtaining such an instrument before the first members stop by.

At first I resolve that I am in a pickle from which there is no practical escape. After giving it some thought, though, I astutely conclude that I do not really need to accurately determine the heights of the chorus members but only have to know their relative heights for the purpose of ranking them. Having tumbled to this conclusion, I go in search of a measuring device. I stumble upon a wooden stick used to prop open a window. By my estimation, the stick appears to be about 1 foot in length.

Through the day, as the members stop by, I measure their heights using my stick. I quickly realize that the stick is not 1 foot long, because the first person's height (a person not much taller than I am) turns out to be 9.5 feet (i.e., sticks) tall. At the end of the day my measurements range from 9.1 to 12.0 feet, and I know these are not accurate measures of the members' height in feet. However, because I have been consistent in my measurements, having used the same stick on each person, I can exactly and correctly assign a place on the stage, according to height, to each member.

More serious real-world problems can be of the same type: those in which the intent is to rank-order entities. Consider the oil well drilling company that has in its portfolio of prospects 150 wells from around the world that might be drilled. Certainly it will be important to eventually determine which wells will yield the greatest amount of oil or gas. However, the first task of the company might be to determine, using a parameter unrelated to resources, which wells have the greatest chance of yielding any resources at all. That is, the company must determine which wells have the highest probability of not being dry holes regardless of the amount of oil and gas they might produce. This approach is taken because expensive tests and data collection and processing procedures are necessary to determine the amount of oil or gas a well might produce. Therefore, the company wants to perform those tests only on wells that they deem to have the best chance of being successful.

While it is true that success is partly defined by the amount of resources discovered in a well, the company is first going to rank the wells based upon parameters not related to resources (i.e., parameters that do not require expensive resource-related tests and data collection and processing tasks). The company deems this to be a measure of technical success. The risk model used to determine a technical success value for each well includes a wide range of input parameters. For example, the model collects from the user a distribution representing the range of probability of political success. This is a measure of the likelihood that governments or partners will agree to terms that will allow the company to proceed with development of the well. Another parameter represented in the model by a distribution is the range in probability that the company might run into objections of sufficient strength from environmental groups. These objections could significantly delay or kill the project.

Other input parameters are of a technical nature such as migration efficiency. This distribution-based parameter measures the likelihood that oil or gas successfully migrated in the subsurface from its deep-seated site of generation to the more shallow location of the trap rock into which the company proposes to drill. If preliminary reconnaissance using geological and geophysical techniques indicates that favorable migration pathways are not likely, then the prospect would receive low probability values for this parameter.

Likewise, the company may, from preliminary geophysical and geological work, evaluate the probability of the existence of an efficient seal for the trap. That is, it is not enough to have oil or gas migrate into potential trap or reservoir rocks. There must exist a relatively impenetrable (to oil and gas) cap rock over the reservoir rock that has prevented, through time, the escape of the oil or gas from the reservoir rock. Based on preliminary geophysical and geological information, risk-model users enter a distribution of values that represent their best estimate for the existence of a sufficient seal for the prospect.

The aforementioned political and technical parameters, along with a host of others, are input to the risk model, which culminates in a distribution of probability-of-success values for each prospect. The methodology for collecting and processing information for each potential well is consistent (or as consistent as reality allows us to be) but not necessarily accurate.

For example, the company may determine that the mean probability of success for Prospect A is 30%. Using their consistent process, the company may conclude that the probability of success for Prospect B is 50%. Therefore, the company deduces that Prospect B is more likely to be technically successful than Prospect A and can rank order these and other prospects. However, the 50% value, for example, for Prospect B does not necessarily indicate that if the company had in their portfolio 100 prospects identical to Prospect B, half of them would be successful. Calibration of such probabilities to the real world often is a practically impossible task. The company does not really know if what they determine to be a 50% probability of success is actually a 25% probability or a 63% probability. For ranking purposes, it does not matter. As long as the company is consistent in their assessment method-ology, and as long as they do not use the resultant percentages as terms in other equations (such as multipliers in equations to determine, for example, the magnitude of the potential resource level from the well), the probabilities resulting from the risk model can practically be used to rank order the prospects.

This process of rank-ordering entities is common practice in many disciplines (see qualitative model and semi-quantitative model examples in Chapter 17). The consis-tency-in-evaluation approach is a methodology that can make practical the solution to problems that at first seem intractable due to problems related to the collection of accurate and precise input data.

LOCKING THE MODEL

Because consistency in assessment is a salient consideration for risk models that produce output values to be used in portfolio-management processes, preventing model users from changing the model is of critical importance. The only way to ensure that risk-model output values can be compared and ranked is to ensure that those parameters consistently have been generated.

It has been my experience, especially when utilizing spreadsheets as part of the risk model, that model users will tinker with the model parameters or equations to get the model to better fit their individual situation. For example, you might have generated a risk model that calculates the net present value (NPV) for a type of project. Your charge is to rank a multitude of projects from all over the world based on the resultant NPV distributions. In January you distribute your model to the offices around the world, and by March you begin to receive the results from the model, which was executed by users in foreign offices. At first you are happy with the results, but after a critical mass of NPV distributions are delivered, your acute business sense indicates that something does not smell quite right.

You request that the individual offices return the model. You find that each office has somewhat modified the model to better suit its particular situation. Most of the modifications are of sufficient magnitude as to make invalid the comparison of NPVs from different offices. Our ability to rank the opportunities has been destroyed.

To avoid this common dilemma, I have found it essential to work with a risk assessment system that allows the party responsible for model-output comparison to lock the model. Locking a model means that the user can input data to the model and execute the model, but does not have the ability to change the input parameters, the number of input parameters, or the equations.

I have found that whenever the issue of locking a model comes up, the groups upon whom the model will be foisted object strenuously because, to them, locking translates to inflexibility and constraint. A locked model need not be inflexible nor constraining. In Chapter 7, I delineate the process by which a comprehensive, flexible, and consensus risk model should be produced. If this prescribed methodology is followed and skillfully executed, the objections to locking the model should be nil.

I find it useful to be able to lock the risk model at two levels, and the risk system I use facilitates this. One level of locking requires that the person locking the model (generally, the model builder) enter a password. In this locking mode, users can peruse the equations that comprise the risk model but are prevented from modifying the equations unless they enter the password. This locking method works well with large and complex models but seems to fall short when employed with simple models. Astute users of relatively simple models will realize that they can simply copy the equations to a new risk model they compose themselves, modify the equations, and send you the results that you believe were generated by your locked equations, but were not. This behavior necessitates a second level of locking a risk model.

A second level of locking models involves preventing the user from any access to the equations and input/output parameter-building mechanisms of the risk-model-building software. When the model is thus locked, users can only see the user interface portion of the model and can only input data, run the model, and review, save, or print the results. Locking models in this fashion can be the only way to ensure consistent assessment.

USERS WILL GAME THE SYSTEM — THE RISK POLICE

In some situations, in spite of best efforts to lock the model and educate potential users, those executing the model will game the system. That is, they will enter values into the system that will, to put it delicately, present their opportunity in the most favorable light.

After implementation of a consistent risk assessment process in an organization, it does not take long for those being subjected to the process to realize just what hurdle value the risk model must produce if the assessed project is to have any chance of being selected for execution. In addition, savvy users quickly deduce which input parameters most significantly impact the risk-model results and thus focus on those input parameters so that their opportunity is perceived best. In this situation it is not long before everyone is living in Garrison Keillor's Lake Woebegone, where all the children are purported to be above average. That is, it is not long before all opportunities exceed the minimum acceptable criteria.

The only mechanism I know of that successfully precludes the input value inflation problem is to establish, for each model type, a cadre of experts that is charged with review of risk-model input. In most organizations this review group is affectionately (sometimes not so affectionately) known as the risk police. In spite of our society's technical and computer prowess, it is this human review that facilitates the successful implementation of a consistent risk process.

For example, within a company it might be the charge for individual groups to present new and potential locations for outlet stores. Groups whose projects are selected will oversee the construction of the outlet. The company cannot afford to build all outlets proposed and thus must rank and prioritize the outlet proposals. This process, necessarily, makes competitors of the outlet proposal groups.

Company management realizes the competitive nature of the groups and, therefore, the propensity to be seen in the best light. To prevent inflationary practices with regard to risk-model input values, the company has established an impartial review board (the risk police). The established risk assessment process in the company requires that before risk-model results for an outlet can be submitted to management for consideration, the input values and resultant distributions must first obtain the approval of the review board. At a risk-model review, it is the job of the review board to do the following:

- Review with the outlet-proposing team the values they have used as input to the model.
- Come to a consensus as to the validity of the input values.
- Act as historical reference and organizational memory. This role allows each new proposal to be judged relative to all past proposals.
- Act as gatekeeper. Proposals that are deemed to be in some way unreasonable should not receive the review board's approval.
- Enforce consistency. One of the most critical responsibilities of the review board is to ensure that all opportunities are consistently assessed so that opportunities can be successfully portfolio-managed.

Clearly, established review boards need to be specific to a risk-model type. That is, the company may need to establish separate review boards for outlets that will primarily sell chemicals and for outlets that will market clothing. Regardless of the number of review boards deemed necessary, review boards should embrace the following principles and should exhibit the following qualities:

- Review-board members should be recognized as experts by the organization
- Review-board members should exhibit impartiality in their assignment

- Review boards should consist of one or more core members for whom assignment to the review board is permanent and is not a rotational assignment
- Review board members should exhibit exemplary interpersonal skills, given that the nature of their assignment is contentious
- The review board should be recognized as being empowered

Establishment of such review boards for each risk-model type can be an expensive and time-consuming task. However, establishment of responsible, impartial, and empowered boards of review is the only risk-process technique I can recommend that will result in realistic and consistent risk-model results.

DATABASING INPUTS AND OUTPUTS

Another element of consistency relates to storing the risk-model input values and results. Although, as mentioned in the previous section, the review boards or risk police can act as an organizational memory, they cannot possibly recall in a risk review all previous input or output values.

If a review-board process has been established, then sanctioned input before project execution and risk-model output values can be stored in a database. The stored values can act as a benchmark for newly evaluated projects. For example, a group proposing a new outlet store might project their sales for a given demographic group to be 10,000 units per month. Values drawn from the database, however, might indicate that sales from similar outlets to similar sections of the population have ranged from 5,000 to 7,500 units per month. The review board might then suggest that unless the group can justify the higher figures, they adjust their sales estimates to match historical values.

It should be noted here the distinction between values before project execution and postappraisal values. In some businesses it is relatively easy to compare pre-execution risk-model input values to those that result from execution of the project. For example, a company that estimates convenience store income for a given geographic location can compare the pre-execution estimates with the actual sales from the store after it is up and running. Both pre- and post-execution values should be stored.

In other businesses, only pre-execution values are generally available. This situation is exemplified by the oil well drilling industry. Companies conducting such business can collect estimates for the probability of a sufficient seal, for a favorable migration pathway for the oil or gas, and for many other parameters. When the well is finally drilled, it either produces economic quantities of oil and gas or it does not. If it does produce sufficient resources, production of the well begins immediately (delayed production is revenue lost) and proponents of the project rarely are afforded the opportunity to verify their pre-drill estimates. If the well is not successful, project proponents are rarely given more funds with which to investigate why the opportunity failed (seen as pouring good money after bad). Therefore, in this industry and others like it, databases contain (primarily) only pre-execution estimates.

NUMBERS THAT REACH DECISION MAKERS

One of the vexing aspects of corporate reality is that there typically exist — even in flattened organizations — multiple layers of management between the pointy end of the spear (business-unit level) at which the moneymaking product production takes place and top of the house executives. Initial projections of production, costs, and revenues are created at the business-unit level. The upward journey of such estimates through management layers affords myriad opportunities for the original estimates to be refined and improved. Often times, if business-unit personnel could see the project-related values that are communicated to ultimate decision makers, business-unit folks would not recognize or be able to relate to the improved assessments and projections.

I have in previous chapters defined the expected value of success (EVS) and the expected value for the portfolio (EVP) and will revisit them again in detail and example in later sections of this book. Each value contributes in a critical manner to the process of decision making. The EVS, though encompassing uncertainty, reflects the range of values — in whatever units — a project might produce if it is executed successfully according to the accepted definition of success. The EVP builds on the EVS by incorporating chances of abject failure. These probabilities are the considerations that would prevent the project from, in a given timeframe, contributing in a fully risk-weighted manner to a portfolio of projects.

Using the EVS and EVP metrics as an example, there exists in repeated and subsequent reviews of a project the temptation to explain away some or all of the uncertainty associated with critical elements and to treat in a similar fashion identified chances of abject failure. For example, a project team might be considering an opportunity that could experience significant delays to schedule due to political wrangling related to granting of permits. Related uncertainty in start dates for the project would have been built into our EVS estimate, providing the delays would not be so dire as to kill the project. Reviewers of the project might attempt to reduce or remove from consideration this uncertainty by claiming that such consideration is double dipping because such delays are already accounted for in other parameters, such as availability of skilled labor, consideration of weather windows, and the like.

If this is true, fine. The questions to ask are as follows:

- When the parameters such as availability of skilled labor were discussed, did the team specifically address the issue of permit delays?
- Are permit delays and availability of skilled labor independent considerations or are they correlated to any significant degree?

If the team did not specifically build in uncertainty related to permit-grant time when considering the availability of skilled labor, or, if they consider the two aspects to be independent or not significantly correlated, then it is not advisable to allow the permit-granting uncertainty to be explained away. Such explaining away processes are deleterious to the overall project assessment process for at least two major reasons:

- EVS and EVP values that truly reflect the project value are less likely to be accurate
- Explaining away the sources of uncertainty or failure denies the project team the chance to consider plans to mitigate the problems

Of the two items listed above, the second one related to mitigation is by far, in my opinion, the more sinister. It does a project and its team a great disservice to explain away problems that might, upon project execution, impact the project's value. Had these problems been specifically considered, such impediments could have had mitigation steps designed to minimize their impact. This opportunity is lost, of course, when such potential hazards are ignored.

So, what is my point here? The basic premise is that the road from project conception to project sanction by executives is a long, arduous path that is replete with opportunities to fine tune the original metrics associated with any investment opportunity. Any proposed risk process should consider this situation and should contain a component that would guard against it. Well, what can be done?

One approach is to implement a peer review process. This mechanism involves, for example, business unit leaders periodically gathering together to critically review one another's projects. This works as long as two diametrically opposing dynamics are not at play. First, if projects are in cut-throat competition for corporate resources, it might be difficult to solicit unbiased reviews for a project from, essentially, competing business unit leaders. Guarding against such biased reviews is possible, but it is perpetually a wrench in the works. This psychological foible can tend to yield project reviews that are overly critical. However, just the opposite result also is possible.

It is with great ease that such peer review processes turn into a good old boys club in which each member knows that he will ultimately be the subject of critical review. The adoption of the group-think philosophy of "I will not hurt you if you will not hurt me" can, almost naturally, spring from both trepidation and true friendships that blossom in such cohorts. This psychological situation, of course, tends to yield somewhat less-than-critical reviews of project viability.

Another seemingly more successful approach to assuring that all risks and uncertainties have been duly considered is to establish some sort of risk police group or process. One form of this approach is discussed in detail in a subsequent chapter of this book, so I will not unduly dwell on the subject here. However, it is well worth mentioning.

A risk police process can take on many forms. One is the creation of a cadre of experts in a discipline or group of disciplines the responsibility of which it is to review all risk/uncertainty-related data for a project. Such groups might be simply bodies that offer review and suggestions or it might be empowered such that sanction from the group is required before a project can proceed. This process might involve a traveling squad of experts that visit project-team sites.

Another incarnation of the risk police process is that of creating a group comprised of experts from disciplines that also comprise the project teams. To distinguish this process from the one described above, this team might be stationary and collect data from project teams via a consistent data-collection form and format. Numbers that pass up the chain might be the responsibility of this review group so they would be inspired to, if need be, have multiple iterations with project teams regarding project-related values before such values could be communicated to decision makers. This process should ensure that all pertinent risks and uncertainties are properly considered.

FORUMS

As already alluded to, gaining consensus regarding risk/uncertainty matters in a large organization is no easy task. Again, contrast the view from finance where risk and volatility (uncertainty) are pursued (low risk, low reward) with the view from a health and safety department that harbors the view that all risks and uncertainties should be eradicated or, at least, reduced to the greatest extent possible. The number of perspectives on risk and uncertainty will be at least as large a magnitude as the number of departments that comprise the corporation.

One approach to attempting to gain consistency in view, to the extent that it can be attained, is to create within the corporation a formal organization that is comprised of practitioners, experts, and interested parties. For lack of a better term, I will here refer to this group as a forum. How formal a function this need be depends much on the culture of the corporation, but some general items of guidance might apply:

- Membership should be as all-encompassing as is practical. Certainly, no critical group or function should be ignored.
- Leadership of the forum should stem from as unbiased a place as is possible. Typically, a corporate representative (i.e., a person not beholden to any specific business entity) who is respected in the field serves best.
- Regular meetings of the forum should be scheduled with an agenda of specific risk-related items to be discussed.
- Such a forum should have top of the house recognition and the ability to have its recommendations influence decision makers and the business units.

It is too much to expect that creation of such a forum will lead to everyone joining hands and singing together, but such organizations have these secondary benefits:

- Promoting cross-discipline communication
- Fostering understanding of different and opposing viewpoints
- Generating consensus recommendations on risk/uncertainty-related items
- Raising the visibility of risk/uncertainty across the corporation and with the decision makers

It is my contention that without the creation of such a forum in a large corporation, there exists little hope of generating universal risk-related processes that make sense for the various and diverse businesses within a corporation. Formation of such a forum is not an ultimate step, but it is one critical element toward being able to successfully implement risk/uncertainty processes in a large organization.

LANGUAGE

In Chapter 2, I include a section on Common Language in which I describe in some detail the problems related to jargon and the benefits of attempting to establish some sort of consensus on the use and definition of risk-related terms. I will not repeat those lessons here, but given that the theme of this chapter is consistency, I would be remiss in my duties if I did not consider the contribution that language can make toward the goal of consistency.

In the previous chapter, I use the example of saying the word "price" to a group of people and considering the various interpretations of the term. Although, as outlined in the section above, it is not realistic to expect that personnel from disparate corporate entities will all subscribe to the same viewpoint on risk and uncertainty, it is of enormous benefit if some agreement can be reached regarding the use and definition of terms.

For example, I might be a member of the forum described in the previous section. I might have a strict definition of the term "risk" that I employ within my discipline and which is accepted and understood by others in this specific field. While that is true, I also recognize that there is great benefit in adhering to a consensus definition of the term when communicating with forum members and others outside my area. So, although I might not agree that the consensus definition of risk is strictly applicable within my cloistered world, because there exists a consensus definition of the term, I can communicate succinctly with other parts of the organization and can clearly understand communications from across the corporation.

So, this is a have your cake and eat it too solution. It allows people to retain their discipline-specific definitions of terms when carrying on everyday work, but also facilitates communication across disciplines. This is not a perfect solution, but in the muddled world of risk and uncertainty which embraces nearly all aspects of a corporation's business, it is a great step forward if properly implemented.

RISK AS A CORPORATE FUNCTION

Until now, and, in fact, for the remainder of this book, I have considered and will consider risk to be a component of multiple disciplines within a corporation. For example, there is risk and uncertainty in finance and in law and in health and safety and in security and so on. This approach is built on the premise that each corporate entity will somehow internally deal with risk and, through employment of a common language, forums, and other devices, will attempt to cooperate and communicate with other parts of the corporation.

An alternative is to make risk a corporate function just like finance or law or health and safety or security. This is typically viewed as a relatively draconian approach, but is one that can yield great potential in a diversified corporation.

Under such a scheme, the risk group would be headed, usually, by a vice president (VP) and the group would enjoy all of the benefits of equal status with other departments. Just as finance might generate rules to which every business must adhere, and just like health and safety creates guidelines that must be followed in myriad situations, the risk function would create guidelines, recommendations, edicts, and tools that are to be utilized by all businesses in the corporation. The complexity of implementing such a function and the effort that will be required should not be underestimated.

For the U.S. government, it is difficult to pass laws that make sense for such a large and diverse country. A 55 mile per hour speed limit might make sense in a major metropolitan area on the East Coast, but in Wyoming you are likely not going to live long enough to get from point A to point B at such a speed. So it is with risk processes and guidelines. Each discipline has to be at least considered and separate sets of guidelines, for example, might have to be issued to accommodate corporate diversity. This can be time consuming, expensive, and difficult to enforce.

On the other hand, this approach is one way to attempt to ensure consistency of application of risk/uncertainty-related processes and techniques across the corporation. It also is a way to bring the importance of risk/uncertainty to the attention of decision makers and business-unit personnel and should keep such considerations from being also ran items at the end of an otherwise structured assessment of an opportunity.

SELECTED READINGS

Campbell, J. M., *Analyzing and Managing Risky Investments, 1ˢᵗ Edition*, John M. Campbell, Norman, OK, 2001.

Koller, G. R., *Risk Modeling for Determining Value and Decision Making*, Chapman & Hall/CRC Press, Boca Raton, FL, 2000.

Schuyler, J. R., *Risk and Decision Analysis in Projects*, Project Management Institute, Newton Square, PA, 2001.

Wysocki, R. K. and McGary, R., *Effective Project Management, 3ʳᵈ Edition*, John Wiley & Sons, Hoboken, NJ, 2003.

6 Communication

COMMUNICATION — JUST ONE SHOT

Classes certainly play a salient role in the communication process; however, classes only serve as a way to preach to the choir. That is, lectures to a class are lectures to those who either already believe or who want to believe. A large part of risk communication takes place between the purveyors of risk technology and those who are skeptical or downright hostile, or those whose positions are buffeted by the results of the risk assessment. Excellent communication skills and processes are crucial.

While in the throes of assessing a situation's risk, you may well have the luxury of time — enough time to schedule critical meetings, enough time to make decisions, enough time to gather data, enough time to build and test the model, and enough time to interpret the results. Relative to these time lines, the risk-communication task is a point source. It has many times been my experience that the results of months or years of effort had to be communicated in a single meeting — usually in an hour or less. This is typical in the business world.

Even if you are given as much time as you need in a meeting, it is a rare instance indeed that you are given more than one meeting in which to convey the risk-related information. An example is dealing with the public on environmental issues. After the first town hall meeting, local organizations, environmental action committees, the press, your competition, and others are off and running with what you said — or what they think you said. Second chances are scarce, and often ineffective if they do occur.

Given that risk communication must take place in an arena like the aforementioned, it is essential that the person or persons doing the communicating be believable, calm and even-tempered, credible, a good speaker, a better listener, authoritative, and empowered. To be sure, this is a lot to ask of an individual, but it is absolutely essential that the person doing the communicating be capable. Such communication is 50% the message and 50% the messenger. A great message and years of effort can be lost if the message is poorly presented.

COMMUNICATION — WHY BOTHER?

Well, if risk communication is itself so risky, why bother? The communication of risk should be undertaken in almost all circumstances. The most important reason is understanding. For example, in the good old days, lawyers would use terms like pretty good chance and better than average to describe to clients their chance of a favorable outcome or verdict. Using risk technology to assess legal situations, lawyers can now discuss with clients the range of probability associated with each aspect of a litigation and can quantify the probability of a favorable outcome and the impact of that result. In such situations it is not the quantitative answer that is so important;

rather, it is the fact that the risk model parameters and input values become the centerpiece for dialogue with clients. Clients, for the first time, can clearly see what goes into making the decisions and can have meaningful dialogue on the various aspects (see Chapter 17, legal model example).

Another example of the benefit of risk communication is that of dealing with partners or investors. Without a stochastic assessment, your best guess at the cost of a project might be $5MM. When the project is complete, the cost was $6.5MM, and the partners and investors are up in arms because you lied to them about the project's cost. Results from a probabilistic analysis might have shown that the project has a small probability of costing as little as $4MM, but also has a small chance of costing as much as $12MM, with a most-likely cost of about $5MM. With this information, investors and partners can, prior to joining the project team, decide whether or not they wish to expose their companies to the possibility, though small, of spending as much as $12MM. If they do decide to invest, they will find the $6.5MM figure well within their comfort zone and, actually, a relief.

Of course, there also are times when there is no choice concerning whether to communicate risk information. In many companies, if a project is to be approved, it is required that results from a prescribed risk process be presented. Even though a risk presentation may be mandated, good communication skills still are important.

KNOW YOUR CLIENTS

Okay, so I'm convinced that risk communication is important and that I should be good at it. Well, just what should I do to help ensure a good communication effort? Good question.

One thing that significantly aids in risk communication is learning to speak customerese. That is, do your homework and learn the terms and jargon that are familiar to the customer. Of course, it is impossible to become an engineer, for example, if you have not been trained as one. Likewise, having performed a risk assessment will not transform you into an economist. However, it is possible to get hold of materials relating to the subject matter and to the risk assessment and to glean from them at least some of the important concepts and terms.

Knowing too much actually can be detrimental. For example, if you are not an engineer but walk into a den of engineers acting as if it only took a week or so to get a grip on this engineering thing, the warmth of your acceptance by the engineering community will be tepid. Nobody likes a smartypants. It is far better to not appear to know too much, but to make it obvious that you have made an honest effort to learn something about their line of work.

Just as we avoid talking about Uncle Jim's weight and Aunt Milli's wart when the relatives visit, prior to visiting a client it is always a good idea to familiarize yourself with their sensitivities. For example, I have often been called upon to perform a risk assessment on an ambulatory corporate corpse that expected the results of the risk analysis to aid in resuscitating the deceased. In such instances it must be succinctly but delicately pointed out that a risk assessment would be of little value. In other cases it becomes obvious that the request for a risk assessment is an attempt to set right something that has gone dreadfully awry.

In the aforementioned situations, and others of the ilk, it is important to discover some means of diplomatically refusing to perform a risk analysis, or, if an assessment is made, to avoid calling their baby ugly. The risk assessment process often can uncover some rather distasteful and, yes, sometimes unwise business practices and questionable financial practices that may have been going on for quite some time. If I had some universally applicable communication solution to the dilemma of illuminating the problem without embarrassing the clients, I would now be writing that solution. The best advice I can impart is to expect this to happen. Investigate potential clients as thoroughly as is practical so you can be as prepared as possible to handle such situations.

In the risk business as in war, there is no substitute for good intelligence. Each client and situation is unique. As stated at the outset of this communications section, it often is true that results of a complex and lengthy risk analysis are expected to be successfully communicated in a short time in a single meeting.

Given that results need to be communicated correctly the first time, it is often good practice to visit with client representatives well in advance of any presentation solely to ascertain just how information might best be conveyed. Knowing that information should be presented in an industrial or academic motif, knowing that it should avoid or stress technical, financial, or other aspects, or knowing that the presentation should be aimed at a specific individual are all examples of the types of intelligence that can be invaluable. Do not be shy or reticent to solicit explicit information from the client organization. Such investigations should be appreciated by the clients, and the presentation undoubtedly will be better received.

FOCUS ON IMPACT AND MITIGATION

The old adage "Pride cometh before a fall" certainly applies to the realm of risk assessment. The risk process can involve the time-consuming collection of much scientific, engineering, financial, environmental, and other types of data. Individuals who solicit and compile such information are rightfully proud of their areas of expertise. As the coordinator of a risk assessment effort, it is your responsibility to temper the zeal of such individuals and control and mold the presentation.

Most clients are not impressed by an overt infusion of technical details or a blizzard of charts and graphs. Presenters should always be prepared to delineate in reasonable detail just how results were generated. However, a risk presentation should focus on the client-organization impact and mitigation facets of the assessment.

If plots and charts are presented, they should convey clearly the financial impact or other data critical to the client for the purpose of decision making. In addition, if mitigative actions are appropriate, the steps and costs of risk-mitigation efforts should be emphasized.

MITIGATION IS NOT ENOUGH

Actions to mitigate real or perceived problems can take many forms. Training or education programs (they are different) might be employed to offset the probability and impact of operator errors. Physical barriers might be erected in an attempt to reduce the adverse affects of a terrorist attack. Published procedures might be devised and issued to reduce the probability and impact of procedural foibles. Regardless of the action taken, it likely cannot be taken for free.

This piece of advice seems so obvious (to me) that it borders on the ridiculous: include the cost of mitigation in the assessment of the opportunity. Just as ridiculous is the propensity for project teams to agree that they could mitigate a problem, agree on the action to be taken, agree to reduce or remove the impact or probability of the problem, and forget to include in the probabilistic assessment the sometimes steep costs associated with implementing the action that mitigates.

Costs can be measured by several metrics, the most common being money and time. Impacts on schedule can be just as dire as direct effects on cost. Now, we all know that delays in time generally translate to cost. However, in most assessment models, deterministic or probabilistic, it is generally simpler to add costs at the appropriate moments in time than it is to shift in time the many steps associated with a project. Therefore, it has been my experience that schedule slips, although ubiquitous and foreseen, many times are not adequately handled — especially when such schedule adjustments are due to the implementation of some action to mitigate a problem.

It is an interesting psychological dynamic that time delays associated with a problem seem to be more likely to be considered than delays associated with mitigation of a schedule-related problem. For example, a team might recognize that the project might be delayed due to lack of skilled labor. It seems typical that they build in a schedule delay to account for the time it could take to obtain such help. A team might decide that they can offer training classes to create skilled laborers from the pool of local talent. Such training will require money and will take time. It is not uncommon for a team to agree on the mitigation action to be taken (train local workers), and having decided on the mitigation step, reduce the schedule and cost impact associated with the problem without considering in the assessment model the schedule delay and cost associated with the actions to mitigate. In any assessment, be sure to offset the benefits of a mitigation action with the cost of implementing the step. It should be the responsibility of the risk-process facilitator to communicate the importance of this process.

For yet another reason, it certainly can be difficult to get mitigation costs included in an initial project assessment. Project teams usually experience significant pressure to reduce costs and improve schedule. It is, therefore, relatively popular to consider actions to mitigate problems and to include the cost savings associated with reduction of the problem's impact and/or probability of occurrence. Because of the pressure to reduce costs and improve schedule, it is equally unpopular to reverse, even partially, the aforementioned benefits by adding back into the analysis the costs and schedule delays associated with the action to mitigate. If a realistic assessment is to be had, such costs and delays must be adequately considered. Again, communication of the importance of this process should not be underestimated.

Earlier in this book, I addressed the problems associated with the check the box mentality, and it is worth revisiting this phenomenon with regard to mitigation. Often, project teams do a credible job, early in the project's life, of considering and recording potential problems, addressing potential mitigation actions, identifying responsible parties, and recognizing costs and other items in, for example, a document like a risk register (see Chapter 10 for an explanation of a risk register). Often, however, that is the end of it.

It is simply not enough to make lists of risks and uncertainties and assign responsibilities to mitigate and otherwise address them. It has to be somebody's job to track progress and to revisit the list as often as is appropriate. This is a difficult thing to implement. Sometimes, the person assigned to track progress is of a lower rank than some of those who will have to be chastised regarding lack of progress. Also, whether or not progress is being made can be exceedingly difficult to ascertain. To determine whether or not there has been progress, the overseer (the person responsible for tracking progress) might have to depend on anecdotal evidence or testimonials from those he is evaluating. This is a tough position, but one that is necessary. One remedy to this set of maladies is to assign overseer duties to someone of sufficient rank and influence. This is the only practical way, in my opinion, to successfully consistently coerce from people valid and reasonably unbiased information regarding mitigation progress.

Well, for those of you who are front-to-back readers of books, glance back to the section of this book in which I offered my universal definition of risk and how it relates to uncertainty. Remember that I defined risk as a pertinent event for which there is a textual description. For example, there might be a risk of an earthquake. The risk, therefore, is simply that there might be an earthquake. This is only a risk if the consequence of the earthquake is dire enough or that the probability of the earthquake is great enough (enough is different for each situation), or both. An earthquake, therefore, might not be on our list of risks if there is a significant probability of such an event but the event is going to happen in a completely unpopulated area such as a desert. Conversely, an earthquake might not be considered a risk if there is an exceedingly low probability that it will impact a populated area. If no mitigation action is considered necessary, then the risk might disappear from our list of risks.

Now, let us suppose that an event is, in fact, considered a risk and does make our list. If it is on our list, then we might try to implement actions to mitigate the risk. It is on the list because we think there is a sufficient probability or impact (or both) associated with the event. So, we might determine what action should be taken to mitigate the risk and assign responsibility to some individual or group. Let us also suppose that, over time, the action to mitigate the risk is successfully executed and either the probability or the consequence (or both) has been reduced. In the real world, it is relatively common that one cannot reduce the probability of a risk to zero or the consequence to zero. So, how low a probability or consequence is required to remove the risk from consideration?

There is no universal answer to this question. Those of you familiar with utility theory (a subject that is beyond the ken of this book) know that there exists a risk threshold that is unique to the individual or group and to the situation. That being said, it is important at the outset to discuss to what degree uncertainties will have to be reduced before a risk will be considered to have been mitigated. There is some danger in this practice, because sometimes targets set are targets attained. (If I had not believed it, I would not have seen it.) However, it is a necessary step to determine at what point a risk is no longer a risk. If a group continues to spend time and money to mitigate a risk beyond the point of diminishing return, much capital can be wasted.

One last comment about mitigation and communication. Sometimes the risk facilitator can be put in an uncomfortable position by teams attempting to explain away risks and uncertainties to which they originally subscribed. This subject is detailed in a previous chapter, but bears mentioning here. For example, a team might agree at the outset that customer commitment related to location might be a sales retardant for a given product. So, built into the model is a factor that limits sales in certain demographic areas. When the model is run and the results do not support the team's optimistic expectations, there can be pressure brought upon the facilitator to delete or significantly reduce the impact of loss of location-related customer commitment. Teams might request that this impact be ignored because we will take account of this in our sales-projection figures. Well, maybe they will.

One element of communication for the facilitator to consider seriously is to communicate to the teams that problems identified and agreed upon in the initial stages of the assessment process will not be explained away later. Such explaining away of impacts until the result matches expectations puts the facilitator and model builder in a precarious position. Teams might use the improved output from the analysis to support their position and, if the facilitator or model builder is of some renown, use the facilitator's name to further bolster their position. So, to head off this problem, clear communication from the facilitator to the team is essential regarding how a risk/uncertainty might be dropped from consideration.

KILLING PROJECTS

Winning the lottery — that is hard to do. (Actually, it is easy; it just is not very likely.) Getting your kids to act as you would have them act — that, in Glenn's world, is nearly impossible. With regard to difficulty, however, these things would take a backseat to the arduous task of killing a project, in some corporations, that has been funded and is underway.

Risk-assessment and portfolio-management processes usually are implemented with the goal of generating information that will aid in the decision-making process. Decisions typically mean choosing one option over another. In a business-unit or corporate portfolio, this can mean selecting projects to execute and passing on others. This, of course, assumes that analyses are being performed prior to funding and staffing of the projects (options).

Many times, however, risk analyses and other evaluation techniques are built into stage-gate processes through which projects pass. A stage-gate process usually is enacted ostensibly to review the project as it progresses to determine whether or not it is meeting its predetermined goals. Having observed many industries, I have formed the considered opinion that once a project has been funded and staffed and has begun to travel down the stage-gate route, the probability of canceling the project is very small. Holistic and independently executed risk assessments can and do run afoul of this tendency.

If a risk analysis at a stage-gate is in fact holistic and if data for that analysis has been collected in an unbiased manner (usually, this requires the involvement of a risk-analysis facilitator who is not part of the project team), such an assessment can clearly indicate that the project is not worthy of proceeding. Okay, now what? In most corporations, there ensues a scramble to fix the uncovered problems which, in many cases, is throwing good money after bad and akin to trying to revive a corpse. That is not to say that some projects cannot be revived, but it is not typically so.

Well, what does this have to do with communication? Risk-process facilitators usually do not relish wasting their time. If the process is going to be that once a project is started and has progressed along the stage-gate route it is very unlikely to be killed, then, I submit, the corporation is basically wasting its time by feigning desire to utilize the results of a risk analysis. One of the critical elements of communication that has to take place between management who promote the use of risk analyses and the business-unit personnel is that projects already underway can, in fact, be killed. And they cannot just say it, they have to mean it. This, in most situations, will reflect a salient change in corporate culture.

If this type of process — that of being willing and able to kill already-started projects — is adopted, business-unit personnel will always feel as though they are on the bubble — and they will be right. The attitude of management might be that this will spur project teams to work harder and smarter, but implementation of such a policy can have just the opposite impact — that of disheartenment and embracing the attitude that it is hopeless. To counter these negative reactions, a corporation must have in place a rational means of dealing with project personnel whose projects are, in fact, killed after inception. The following section deals with this dilemma.

DIPLOMATICALLY HANDLE WINNERS AND LOSERS

The risk assessment process generally serves as a prelude to portfolio management. The portfolio management process, for all of its stellar qualities, does bring to the organization the dark side of designating winners and losers. In the majority of situations, a risk analysis is performed to aid in making a decision. Such decisions typically involve selecting one option or, to wax philosophical, to take one road rather than another. The proponents of the road not taken must be dealt with carefully.

As the ancillary risk assessment person, you may well find yourself in the position of having to explain to an organization the results of the analysis. Results may indicate that some members of the audience are no longer going to be involved in the assessed project. You most certainly will not be empowered to offer a place on the winning team to those whose proposals were rejected. This situation can be volatile, and deft and delicate handling is required.

As the presenter of risk assessment results, you are not empowered to offer organizational solace to those who find disfavor with the outcome. The only recourse is to emphasize the logic behind the selections made (and you had better have some) and how the decisions taken best address the needs of the greater organization. This sometimes is a difficult or impossible message to successfully convey. The best communications efforts are comprised of the following points:

- The decision that was made
- The logic and process leading to the decision
- The benefit to the overall organization

A discussion of how other decisions would have been worse should be avoided. Certainly it is not beneficial to get pulled into a discussion as to why and how the things that were not selected are worse than the things that were selected. Diplomatic and even-tempered elucidation of the three points listed above is the best communication solution to this difficult but often-occurring communication dilemma.

COMMUNICATING THE VALUE OF RISK ASSESSMENT

In the business world, if a technology is to survive, the value of that technology must be succinctly communicated. Especially in the business world, it is a particularly vexing task to prove that a technology adds value (i.e., makes or saves money). In the academic arena, if practitioners understand, accept, and use a technology, then that technology is deemed a success. In the scientific world, if a technology accomplishes a heretofore impossible or impractical chore, then the new method is proclaimed triumphant. In the business area, however, it is the bottom line that counts.

It is a relatively simple matter to calculate the cost of a given technology: "Let me think, I have four people working on this, plus materials, plus computer charges, plus travel expenses, plus...." It is a much more arduous undertaking to document the (dollar) benefit of a technology. So it is with risk assessment.

Those pressed to document the cost/benefit ratio of a technology such as risk assessment typically attempt to associate some dollar value to use of the technology. Do not even bother to go there. Generally, the technology provider is not the primary technology user. Therefore, the provider has to request from the user what portion of the user's financial success can be ascribed to the use of the technology.

In most modern business environments, technology users, in one way or another, are charged for the privilege of using the technology. Therefore, when the technology provider requests to share in the riches of success, the technology user sees that request as akin to the seller of a spoon wanting to take partial credit for the culinary extravaganza created by a chef. "We paid you for the spoon — what else do you want?"

Another time-honored roadblock to associating positive value with technology use is the argument, "If you share in the success, then you must share in the failures." Consider the situation in which a new treatment for a fatal disease is successful in preserving the lives of 30% of the patients upon whom it is used. Well, 30% survive, but 70% die (no matter that they would have died anyway). Should you not share in the cost of the failures too?

Another example of this same phenomenon is evident in the world of exploration oil well drilling in which the failure rate is almost always greater than 50% (i.e., more than 50% of the wildcat wells drilled in the world tend to be economic failures). So, suppose we have developed a new seismic technique that is applied to virtually all prospects, and it raises the economic success rate from 40 to 45% in wildcat drilling. Does the new technology only get credited with some dollar portion of the 5% increase, or does it have to share the dollar burden from all those unsuccessful prospects to which it was applied? This generally is a conundrum that remains unresolved.

I have, however, saved the best for last. No one can know the results from the road not taken. Therefore, if a venture that used the new technology is successful, would it not have been successful anyway? Suppose you urge me to wear a special hat while popping popcorn. I take your advice, and each time I attempt to pop corn, it works. This results in a 100% correlation. Is there really any connection? How do you, as the technology provider, respond when the technology user claims that they would have been successful, or would have made that good decision anyway (even without your stuff)? In businesses that are keenly eyeing the bottom line and in which the technology provider has been paid for the use of the technology in the first place, the provider should not be aghast at this tenor of response.

So, if the longevity of a technology depends upon the promoter proving and communicating the value of the technology, just how do you do that? Many years of beating this (apparently dead) horse has led me to draw the conclusion that you have to prove worth indirectly — not with dollars. One metric to consider and to communicate to those holding the purse strings is the amount of repeat business you enjoy. The real measure of success is not in getting new customers — someone can always be suckered into trying it once — but is in the number of people or businesses that time and again use the technology you provide. Written testimonials from customers also are helpful.

Yet another technique I employ is to request that users of the technology put the methodology on a list of techniques and processes they used in their venture. I ask them to list the technologies or processes in rank order (no scale) with the perceived most useful item at the top of the list. The technology or process thought to be least beneficial should occupy the bottom position. Lists from various users rarely will have many items in common. If, however, your technology consistently ranks near the top of the individual lists, this fact should be prominently communicated to those who have the power to eliminate unnecessary items when budget-cutting time comes around.

If I were you, I would be asking myself why I, the reader, was being subjected to this decidedly melancholy diatribe about the near futility of associating value with technology. The answer is because risk assessment is among those techniques to which these anecdotes apply. Typically, a risk assessment forces quantification, documentation, and rigorous integration of elements of a venture. These elements would otherwise have been evaluated and combined by less arduous, seat of the pants methods. To adduce that, without risk assessment, risk users would not have made the decisions they ultimately embraced is a claim that usually cannot be substantiated. Therefore, like so many technologies, it generally is clear how much cost, time, and effort were required to implement a risk assessment process, but much less evident is the benefit of invoking risk technology.

Risk processes established in organizations do not have a life force of their own. Without constant care, feeding, and promotion, such processes tend to fall by the wayside if taken for granted — only to be resurrected when it ultimately is realized that what was discarded was an essential element of the business. If risk assessment or an established risk process is to survive or flourish, the implementation of metrics and the communication of the compiled supporting data are crucial.

IT STARTS AT THE TOP

Just because a CEO or president sits at or near the top of a corporation does not mean that such a person can wave his hand and enact sweeping cultural or procedural changes in a corporation. In many instances, people at the top of the house find it unnervingly difficult to enact cultural or process changes without concomitantly destroying the positive aspects of current practices. The attitude of subordinates that the power to make sweeping changes in an organization sits, for example, with the CEO, is one of naivety. Frustration with making changes can and does affect the top of the house just as much or more than it infects those below.

One change-enacting approach available to management is that of communicating expectations. For example, people at the top of the house might find it difficult to issue a specific dictate that can be universally embraced and implemented in the multitude of disparate arms of the corporation. I hearken back again to the 55 mile per hour speed limit set by he U.S. government. Such a rule might seem reasonable in a major metropolitan area, but in Wyoming …

Communication of an expectation can circumvent many of the problems associated will the issuance of a specific edict. With regard to risk processes, it might be communicated to the corporation that it is expected that:

> Any project coming before a board of review will have had an appropriate and holistic risk analysis performed on that project. Such an analysis will highlight the uncertainties and chances of failure associated with the project as well as the upside opportunities and potential actions for mitigation.

Or some such words. Such a communication of expectations does not tie a project to a specific type of analysis, nor does it preclude the application of discipline-specific practices in a risk assessment. This type of communication plays into the hands of anyone attempting to implement a risk culture in a corporation because such a change cannot be successfully enacted without support from top management.

Quite simply, if nobody is asking for this, then the promoter of a risk-based culture will find it exceedingly difficult to persuade business-unit personnel to spend time and money in implementing a risk process the results of which have no takers. In the past, I have been guilty of first training the practitioners in the skills of risk assessment and risk-model building. It was my general experience that when I visited those same folks sometime later, they had stopped creating risk assessments the results of which would be used by anyone other than themselves. It turns out that if management personnel above the practitioners do not want, understand, or appreciate the risk-model output, then such information is used, at best, only by the practitioner. Such practitioners might, in fact, be chastised for wasting their time on such pursuits.

So, the bottom line is that if cultural and process changes such as implementing a risk-based culture are not communicated from the top, then those changes have very little chance of succeeding. Communication from the top should be carried out in such a way as to allow sufficient flexibility to accommodate business-unit differences while being specific enough to garner the results required. Without support from the top, there is a significantly diminished probability of success.

CREATE SUCCESS STORY PRESENTATION

Okay, so if you buy what I am selling in the previous section, you take to heart that upper management has to communicate that risk processes are to be associated with each project. Well, just how does one convince top management that they should issue such a communication?

As stated previously, decision makers are not necessarily interested in a more cool way to make calculations — such as through a probabilistic analysis resulting in ranges, probabilities, and the like. *What they want is a method the results of which will allow them to make better decisions.*

You can never know the results of the road not taken. That is, it is very difficult to convince someone that just because you had them go through the process of performing a risk analysis that the result is better than what they would have gotten had they not done the risk analysis at all. In fact, I have been told to my face that had I not bothered the business unit with this risk stuff, they would have arrived at the same place, in the end, and would have saved all the time and money that it took to implement the risk process.

To sell the risk process, it is futile to make the argument that, for example, the financial or production values are better than those that would have resulted sans the risk process. It might be true — and I believe it is — but given that you cannot compare the results with the road not taken (the road of no risk process), such an argument rings hollow.

The communication to decision makers should not focus on better numbers but should promote the improvements in project assessment and business practices. For example, one of the salient aspects of the risk process is, early in the project life, to brainstorm about the potential upside and downside aspects of the project. Such a practice can promote early implementation of actions of mitigation and can hone the scope of the project and focus spending most effectively. In addition, recognition of the ranges for project-specific parameters can indicate where most uncertainty lies which should spur appropriate actions to reduce it. Creation of ranges for output parameters can serve as benchmarks and reality checks for target values generated by the business (i.e., where does the business-generated output expectation fall in the range of things likely to happen?).

It is difficult to argue that a process that requires early identification of problems, recognizes uncertainty, and promotes actions of mitigation and focused spending is not one that benefits the business. These are the aspects of the risk analysis that impact decision making and upon which the process should be sold.

To make this argument to upper management, the promoter of such a change should be prepared to demonstrate how implementation of such a process actually impacted the decision-making in a real project. This, of course, means that the risk-promoter has had to convince at least one project to give this a try — even though upper management has not asked the project to do so and likely will not yet appreciate the time, effort, and money needed to generate results that they currently have no means of utilizing. This, of course, also means that the promoter of such a process is capable of doing a credible job of enacting such a risk analysis on the project that presents itself as willing to be a guinea pig.

This is, admittedly, the classic chicken and egg problem. One has difficulty getting management to issue a communication of expectations without management being convinced of its worthiness and one also has difficulty convincing a business unit to spend time and money on a risk process that upper management does not yet appreciate. The only practical means of breaking this logic circle is to identify some project that is willing to give it a try. This often means calling on project leaders who are in charge of projects with which the risk promoter is most familiar. This also means that the risk promoter has to have sanction to spend time and budget on such an experiment. These are the practical aspects and realities of attempting to influence the corporation.

This can be an arduous task, but do not give up. It is only through persistence that such a process is successfully implemented. I have been witness to successful implementations of corporate risk processes in many different corporations, and it has been those who persisted in the pursuit who have been successful.

WEB STUFF

This last section is one that I offer with enormous reluctance. At the time of this writing, Web pages are all the rage. That is great. I am actually typing this book using modern word-processing software and am a convert to most computer-based processes that affect my life. Okay, that said, I have a few reservations that I feel I must relate with regard to the use of Web pages.

First, Web pages are great tools for use as repositories for information. An interactive Web page can collect and relate information to many individuals who, in the past, might have had to be addressed individually by a person. This is a great efficiency. However, a Web page is only as good as its content.

I am not against the use of Web pages, but I do rail against the establishment of such a page as a universal remedy for complex problems. First, I do not know too many business-unit personnel — especially those at higher levels — who routinely scour the intranet for new information. Most people are exceedingly busy and distracted, and it is a rare occasion that they would consume any free time they might have with a Web-surfing exercise. Again, I am not against the use of Web pages, but it does amaze me that some people think that because they have put their two cents up on a Web page that they have communicated with the corporation and that other more traditional methods of communication need not be employed. Nothing could be farther from the truth.

Although much of the information related to risk and uncertainty can be at least outlined on a Web page, it is difficult to convince people — even those interested in the subject — to visit the page. This is doubly true if the page needs to be repeatedly visited for updates, latest news, and so on. Making a Web page a destination for practitioners and decision makers is just plain difficult.

Web pages and their content are great tools, but they should be combined with more pedestrian methods of communication such as meetings, personal communication, the teaching of classes, internal advertisement of various sorts, and so on. It is my view that a Web page is not an end in itself, and that much effort not at the keyboard is required — combined with Web-page content — to implement a risk/uncertainty process successfully in any large organization.

SELECTED READINGS

Argenti, P. A., *Corporate Communication, 2nd Edition*, McGraw-Hill/Irwin, New York, NY, 1997.

Atkins, S., *The Name of Your Game — Four Game Plans for Success at Home and Work*, Ellis & Stewart, Beverly Hills, CA, 1993.

Block, P., *The Empowered Manager: Positive Political Skills at Work*, Jossey-Bass, San Francisco, CA, 1990

Jasanoff, S., Bridging the two cultures of risk analysis, *Risk Analysis 2*, 123–129, 1993.

Johnson, B., *Polarity Management — Identifying and Managing Unsolvable Problems*, HRD Press, Amherst, MA, 1992.

Maister, D. H., *Managing the Professional Service Firm*, The Free Press, New York, NY, 1993.

Morgan, G. M. and Henrion, M., *Uncertainty — A Guide to Dealing with Uncertainty in Quantitative Risk and Policy Analysis*, Cambridge University Press, Cambridge, MA, 1992.

Pasmore, W. A., *Creating Strategic Change — Designing the Flexible, High-Performing Organization*, John Wiley & Sons, New York, NY, 1994.

Roussel, P. A., Saad, K. N., and Erickson, T. J., *Third Generation R&D — Managing the Link to Corporate Strategy*, Harvard Business School Press, Boston, MA, 1991.

7 Building a Consensus Model

WHAT IS THE QUESTION? — MOST OF THE TIME AND EFFORT

It happens almost every time. Someone from a group in the organization will call me or send me email and request help with a risk assessment model or process. When I arrive on the scene, typically there is a conference room full of people to greet me. Folks from all walks of life are gathered — managers, political specialists, economists, engineers, other scientists, environmental specialists, legal counsel, and others. I liken this gathering to the three blindfolded persons and the elephant. If you were to ask these three persons: "Do you have an elephant?" their universal answer would be "Yes, we certainly do." However, if you asked each person to describe the beast, the one holding the tail would say it is like a rope, the one touching the leg would indicate that it was like a tree trunk, and the one holding the ear would relate that it was like a big flat thing.

Initial meetings with groups attempting to build a consensus risk assessment share some characteristics with the three blindfolded individuals. If you ask: "Do we need a risk assessment?" the universal answer likely will be "Yes, we do." However, if you ask: "Who thinks they can state in one sentence (two at the most) just why we are here today?" many hands in the room will go up. Each person has their idea of just what question is to be answered. I can tell you from experience that there will be little commonality to their answers. It is an important function of the facilitator to attempt to meld the diversity of opinions and view-points into a risk model — a consensus risk model — that everyone agrees is sufficiently comprehensive, rigorous, and flexible to accomplish the task at hand. This is a tall order.

The very first thing to do when beginning to facilitate such a group is to ask someone (who is not the boss) to write on a flip chart or board, in a single sentence, just what they believe the problem to be. When this has been done, ask the rest of the group if they agree with exactly what has been written. If they have no other opinion (which they will not, if you let the boss write first), you will be in the presence of a very unusual group. Typically, engineers see the question as having to do with the engineering aspects of the problem. Economists likewise see it as fundamentally an economic argument. And so it goes with each person and each discipline (i.e., what does the elephant look like?). Arriving at a consensus as to just what is the problem (or problems) that is going to be solved is a tricky proposition.

CONSENSUS MODEL

When multiple projects of a similar type need to be evaluated (like the aforementioned plant-construction projects), a consensus risk model should be constructed with the aim of applying this model consistently to all projects of that type. For details on how to define and build type models, see the related section of Chapter 5 of this book.

The ability to rank, compare, and portfolio-manage risk-assessed opportunities is predicated on the assumption that the assessed opportunities can, in fact, be compared. Comparison of opportunities requires that all opportunities be compared on the basis of one or more common measures. Typically, these measures are financial, such as net present value, internal rate of return, and others. However, measures that are not financial also are common. Measures such as probability of success (technical success, economic success, etc.), product production, political advantage, and others can be used to compare opportunities.

For example, let us consider the situation in which a business manager is charged with the task of deciding which of several competing product lines should be selected for marketing. The product lines may be diverse. One line may be footwear, one hunting gear, one basketball equipment, and so on. The manager may decide to compare these opportunities on the basis of net present value to the company. The challenge, however, is to decide whether to allow the departments for the product lines to construct their own assessment models or to attempt to generate a consensus model.

If the comparison of competing product lines is likely to be a unique or rare event, then the manager might decide to allow each department to devise its own assessment model as long as the result of the analysis is the net present value of that product line. If this is the path taken, it will then be up to the manager to be sure that all critical elements (taxes, labor costs, other expenses) are considered in each department model. Such an undertaking can be a daunting task. This method of comparison may be most efficient in the case of assessing events that are not likely to occur again.

The manager, however, may decide that he will many times be confronted with making such decisions. In this case, it likely is more efficient to attempt to build a consensus model which can be applied to the three product lines presently being considered and to all other product lines that may be considered in the future. A graphical example of these two contending philosophies is shown in Figure 7.1. It should be noted that to build a risk model capable of considering all the intricacies of a multitude of product lines, the manager will have to assemble critical representatives from each department and engage them in the risk-model-building process. This can be a time consuming and sometimes a nearly politically intractable task. However, the effort will pay off with a risk model that can be applied repeatedly to new opportunities or sets of opportunities. Those opportunities can then be compared with options already assessed. Portfolio management is the result.

It should be expected that the risk model will be an evolving entity. With the repeated application of the model to new and slightly different situations, the model will grow and evolve into a more robust and comprehensive assessment tool. Appro-

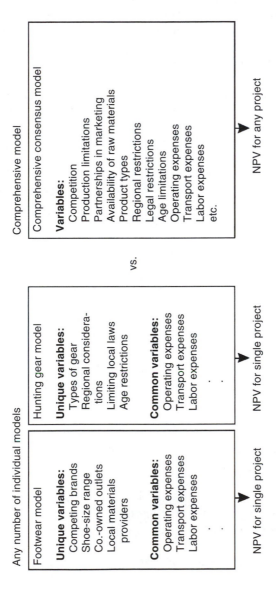

FIGURE 7.1 Example of plans for multiple individual models versus a consensus model.

priate resources for the maintenance and enhancement of such a model should be considered essential from the beginning of this process.

GROUP DYNAMICS

The real trick in arriving at the consensus solution is to do it without hurting anyone's feelings or having someone feel as though his contributions are not being justly considered. There are many approaches used to resolve this type of conundrum; however, they all share the common thread of orchestrating group dynamics.

The first piece of advice I generally give to a group attempting to reach consensus on a question or model is to use a facilitator (a discussion or model-building leader) who does not have a stake in the outcome of the project. In fact, the best facilitators are those individuals who

- Have good stage presence
- Have good interpersonal skills
- Understand the process of orchestrating group dynamics in the risk-model-building arena
- Completely understand risk technologies

In fact, the best facilitators are those who exhibit the attributes listed above and who have little specific knowledge concerning the problem to be solved. Relative ignorance not only imparts an aura of impartiality, but allows the facilitator to ask fundamental questions that otherwise may be ignored or the answers to which would be taken for granted. Discussion of fundamentals many times can be of great importance in the process of reaching consensus.

For example, when there is a gathering of individuals concerning a project for which a risk model is to be generated, many of the individuals know one another. This can be both beneficial and detrimental. The benefit is derived from the feeling of comfort that comes with familiarity. This comfort factor, shared by the people in the room, can be used to move things along because introductions are not necessary and meeting new people jitters is not a factor. Contributions to the discussion are more freely given in familiar surroundings. However, familiarity also can work against you.

In many situations, people who know one another carry baggage with regard to past relationships. In addition, there may be competing entities in the room. This may take the form of competing individuals (only one engineer is going to get to work on this), competing groups, or other forms of competition. This and other things such as downright bad blood have to be considered when attempting to facilitate a risk-model-building exercise and must be mitigated in a constructive manner. There are many group dynamic processes that can help.

WRITE IT DOWN

After the ice has been broken by the first person expressing his written opinion as to just what is the problem, the facilitator should ask those who take issue with the written statement to write on a piece of paper, in a single sentence, just what they think is the problem. If the facilitator allows verbal expression of the problem(s), chaos results because, invariably, an argument or heated discussion ensues. The next facilitation steps are as follows:

- One at a time, have volunteers come to the board and write down their expressions of the problem
- Following each statement, direct the group to find some common elements in previous statements and the newest one
- Keep a separate written list of the common elements in full view

In Figure 7.2 is depicted a table that might result from the process described above. In the example shown in Figure 7.2, we are really lucky that a common thread concerning the economics of the project could be found in all points of view. To calculate the net present value (NPV) of the project, we will of course consider the engineering, environmental, economic, political, and other parameters, but at least we now have a common and agreed-upon goal for the risk model.

If a reasonable list of common elements has been compiled, then the next step is to engage the group in attempting to generate a concise (i.e., short) statement that captures the salient common points. This process does not have to result in a single statement. More than one problem may be recognized; however, bear in mind that separate problems may require separate risk assessments. It is the facilitator's job to be sure that all contributing parties in such a process are recognized and that all ideas are given a fair hearing. In this case, we may settle on the statement, "What is the cash-flow situation and NPV of a selected plant expansion scenario?"

PLANT-EXPANSION MEETING — JUST WHAT IS THE QUESTION?

Individual Statements of Problem

- Engineer 1 — What is the best affordable construction method for the plant-expansion scenario selected?
- Engineer 2 — What is the plant-expansion scenario that yields the highest net present value?
- Economist — Regardless of the plant-expansion scenario (i.e., technical plant-expansion details), what is the net present value of expansion of the product line?
- Environmental Engineer — What effect will spills and other environmental concerns have on the profitability of the project?
- Political Consultant — How will the timing and economics of the project be affected by governmental attitudes and requirements?

Dynamically Kept List of Common Points

- Common points between statements 1-2: Scenario and affordable plant expansion
- Common points between statements 1-2-3: NPV and expansion
- Common points between statement 1-2-3-4: Profitability (i.e, NPV)
- Common points between statements 1-2-3-4-5: Economics (i.e., NPV)

FIGURE 7.2 Compilation of ideas for the purpose of determining just what question the risk assessment will attempt to answer.

If a high-level and general risk model is constructed, then several, if not all, plant-development scenarios will be able to be evaluated with the same model. This is now the challenge.

SORT IT OUT

If no common elements, or very few, are identified (the case when people have *very* different ideas as to just what is the problem), then a sorting mechanism must be used to arrive at the most important of the issues expressed. This sorting can take may forms. A hierarchy diagram (see Selected Readings at the end of the chapter) is a commonly used tool. In a typical hierarchy diagram, the policies are listed at the top, the strategies are listed in the middle, and the tactics at the bottom. The typical form for a hierarchy diagram is a triangle. The sorting mechanism I use is somewhat different. During a break in the meeting, the facilitator should copy all of the sentences and ideas listed on the board to a piece of paper and make a copy of the paper for each participant. When the meeting resumes, each person is asked to label each statement as either a focus item or a not considered now (NCN) item.

Statements labeled as NCNs can include items that we will accept as fact. For example, someone may list, "We must realize returns in excess of our costs." Although this is a good guiding principle, it likely will not be part of the expression of the final question. NCN items can also include details. These are items that may eventually be addressed, but not as part of the primary problem. For example, someone may state, "We must train our people in food-processing regulations." This may be true if we decide to implement the food-processing strategy; however, if we do not implement that strategy, training is not necessary. Statements labeled as focus items are those that we think are the important problems upon which we should focus our attention. These are the statements that will eventually form the essence of the problem to be solved.

After each person has labeled all statements with one of the two labels, the papers are gathered and those statements that were identified as focus statements by more than about half the participants are printed on a new paper and again distributed to the group. If there are relatively few statements that were identified multiple times as focus items, then it is expedient to resort to the common-element process described previously. That is, can we express the elements of the few focus statements as a single problem to which (nearly) everyone can ascribe? If so, then the statement of the problem should be generated from the few focus items. If, however, the number of focus items is great or diverse, then a second round of labeling only the statements previously identified as focus items is undertaken. This time, the people in the group are asked to sharpen their focus and again label each item as either an NCN or a focus item.

In particularly difficult situations, several iterations of this process may be required to arrive at a small set of statements. The small set of statements should either be reduced to a single expression of the problem by identifying common elements in the statements, or should be considered to be separate problems if no common ground can be found. In the latter case, the facilitator should make the group aware that separate risk assessment models may be required for each statement and that the group will have to decide in which order the individual problems will be addressed.

GROUP DYNAMICS AGAIN

Sorting schemes such as the one described here are valuable tools, not only for arriving at consensus with regard to just what is the problem, but for fostering good group dynamics. When an individual in a group contributes an idea or statement, that person would like to feel as though his or her contribution has been given a fair hearing and consideration (no matter how inane the contribution may seem). Using these sorting techniques, the contributed ideas that do not seem to apply need not be singled out as inappropriate.

The sorting process described will discard less useful contributions along with many others of greater merit. In this way, the feelings of people who have contributed tangentially useful ideas are not singled out, nor do they feel criticized for their contributions. It is an important part of the facilitator's job to keep as many people contributing to the process as long as possible. Tools like the sorting mechanism help very much in this regard.

UNITS

Following succinct definition of the problem to be solved, the next item to be decided is just how, in the end, the answer will be expessed. This may seem like a technical detail, but believe me, it is not. It will guide the rest of the entire risk assessment exercise.

Typically, if the question is well defined, a discussion of units will follow. For example, if we are considering the problem of adding production capacity to an existing plant, we may wish to express the final answer in tons of product, or as throughput of material, or as a financial measure such as cash flow, or after-tax income, or net present value, or any of a host of other financial measures. In any event, the units of the answer must be decided upon in the beginning of any risk assessment.

In the next section, I will discuss in detail what I call overarching categories. Examples of such categories may be political, environmental, technical, and other considerations. Each of these categories typically is broken down into numerous variables. For example, categories and variables might be as follows:

- Economic
  - Price (dollars per gallon) of product
  - Advertising costs
  - Litigation costs
  - Construction costs
  - Tax rate
  - Operating expenses
- Environmental

- Probability of a spill
- Attitude of local government toward pipeline construction
- Previously existing environmental damage
- Availability of emergency cleanup equipment
- Technical
  - Capacity of new pipeline
  - Capacity of existing facilities

It is important to establish the units of the answer because all variables that contribute to the solution of that answer must, in the end, be able to be combined such that the result is in the desired units. For example, the answer variable may be after-tax income in dollars. In this case, we must be able to combine in our risk model variables such as probability of a spill, attitude of local government toward pipeline construction, and all other variables in such a way as to result in cash flow in dollars.

So, knowing that the end result of the analysis has to be expressed in dollars, we must keep in mind when discussing the probability of a spill variable that the probability has to be multiplied by some dollar amount. For the variable "attitude of local government toward pipeline construction," we may decide to integrate this into the risk model by asking two questions: What is the probability that the local government will stop the project altogether? What is the per-week cost of delay if the local government holds up construction? This variable might also be handled by implementing a unitless scale that can be translated into dollars. For example, we may wish to express the attitude of local government toward such projects on a scale of 1 to 10, with 1 indicating a good attitude toward such projects and 10 indicating a bad attitude. You may then in your risk model use this 1-to-10 value as a multiplier (or as a pointer to a multiplier) for a value or distribution of values which represent, for example, the weekly cost of local government delays.

It should be clear from the examples listed here that the definition of the output or answer variable and the units in which it will be described are important. This must be known prior to defining categories and variables for the risk model. Certainly it must be taken into consideration when phrasing questions that will be used to prompt information providers.

OVERARCHING CATEGORIES

Every problem is composed of component parts. In the risk business we generally think of these component parts as variables. We do this even though some of the components, in the end, will not vary (constant values or chances of failure, etc.). Thus, I use the term variable here in the loosest sense.

The variables that comprise a risk model can be categorized. In constructing a risk model, I generally approach the situation by first defining what I call overarching categories (OCs). Later I help the group to generate the variables that collectively define each category.

For example, in the case of a plant-construction risk model, I would next challenge the group to define categories for consideration in the risk model. When building a plant, certainly you would have to consider aspects in at least the following categories:

- Political
- Environmental
- Labor (union)
- Technical
- Commercial
- Financial

Probably others will need to be considered as well. When making the list of categories, be sure to generate a definition for each category. For example, when we list labor as a category, just what do we mean? If we are going to consider just the cost of labor, then why would that not be considered under the financial category? Considerations such as the availability of qualified labor, negotiations with local unions, and other labor-related considerations may justify labor as a category separate from the financial one.

Categories are useful for organizing variables and for orchestrating discussion of how the risk model will be constructed. In addition, categories help to organize presentation of the risk model to the end user. When risk models are constructed, variables typically are added to the user interface in an order that generally is dictated by their appearance in equations or by the order in which they arose in discussion of the model. Categorizing the variables allows for a logical grouping and flow of information in the model user interface.

NO SUCH THING AS A UNIVERSAL MODEL

Okay, I have done a lot of talking about type models (Chapter 5) and consensus models — and rightfully so. Models of this type are very useful when a specific problem has been identified and when that problem is shared by multiple and sometimes disparate corporate entities. However, such models should not be viewed as universal in their application.

When type and consensus models are built, they are carefully constructed from extensive interrogation of interested parties regarding a specific problem or problems. These models can be extensively utilized to address the narrowly-defined problem, but should not be misinterpreted to be applicable to a broad spectrum of problem types, or, even to a problem that differs in only minor ways from the situation the model was intended to address.

The perfect analogy is the typical spreadsheet. Engineers, for example, might have constructed a spreadsheet model to estimate costs associated with refurbishing compressors at a production site. Such a model might find application across the site due to similarity of compressors used and the utilization of the equipment. However, engineers at another site — or at the same site where different types of compressors are used or similar equipment is used as an integral part of a different activity — will likely find that the model cannot be successfully applied, without considerable modification, to assess their problem. The point is, just because a dandy consensus model is built, it cannot be expected to be universally applicable — even within a seemingly focused area of interest.

It is a rare occurrence when a sophisticated model can be applied without modification to a problem for which it was not specifically designed. In fact, the more sophisticated the model, the less likely it is to be applicable to a range of assessment types. For example, I have often been asked by those who are tasked with probabilistically assessing a financial investment whether or not I have in the past built a risk model that addresses the type of problem a client is facing. My answer is invariably "yes," but the chance is very small that a model built for some other — but seemingly related — problem could without extensive modification be applied to the new set of circumstances. Because such models are typically imbued with a host of very problem-specific algorithms and processes, I have always found it much more difficult to try to recapture (in my mind) the logic utilized and to change that logic to suit than simply to delineate the new problem in detail and build a new model.

So, although type and consensus models are exceedingly useful in addressing problems that can be similarly defined, such models should not be seen as applicable to problems that differ, even in minor ways, from the situations for which the models were specifically designed. It is the sad but true state of our algorithmic construction and of time pressures of everyday business that lead us to design models that elegantly address a specific set of business dilemmas. Those models rarely find application outside of the narrowly-defined world for which they were constructed.

DIFFICULT TO GET PEOPLE'S TIME TO GO THROUGH ALL THIS

This chapter is just chock-full of suggested steps related to model building. Having employed these practices myself and having here recommended them to you, I can hardly now back away from enthusiastic recommendation. However, a word of caution is in order.

In a perfect world, all of the steps defined in this chapter are salient parts of a well thought-out and executed model-building process. It is, however, not a perfect world. People are busy. Economic pressures are real and acute. It is difficult enough to get a person's or a group's attention, much less to convince that person or group that they should take the time and trouble to enact the long list of steps delineated in this chapter.

The reality is that people are in a hurry. Getting them to listen politely to how risk modeling should be done is a task, but not an insurmountable one. What is nearly insurmountable is cajoling them into seriously considering and applying each of the risk-model-building process steps and recommendations. Nearly every person or group with which I have interfaced takes the attitude: "Yeah, that is great, but I am in a bind here. Are there not some shortcuts we can take?" Of course there are shortcuts, but each piece of the overall process that is abbreviated or skipped does compromise the result. So, how does one entice a person or a group to consider the comprehensive plan?

I have found that one cannot sell the idea of comprehensive risk modeling based on numerical results. Because one cannot know the road not taken, it is difficult to convince someone that, for example, the economic outcome from a business process that included a comprehensive risk analysis is, in fact, better than one that would have resulted had a risk approach not been used at all.

Nothing sells like success. Success has to be defined as the unquestionable benefit gained from implementation of the risk process. For example, it is difficult to argue that it is not a good thing to come to consensus regarding what problem is to be addressed. It is difficult to argue that it is not of great benefit to sleuth out the components of a problem and to engage groups in conversations regarding the range of potential outcomes associated with each component. It is difficult to argue that being able to diagram the problem and address group-dynamic problems is not a time- and money-saving process. And so on.

The best argument for enacting the suggested risk-model-building steps is to relate to the new client the real and tangible benefits that were realized by a previous application of the principles. This, of course, means that you would have to have successfully engaged a group in the list of risk-process activities outlined in this chapter. So, just like in Hollywood, it all starts with getting that one big break. It might take time and be a road of considerable frustration, but finding and working with at least one corporate entity that embraces and successfully applies the suggested process is worth the rocky road.

SELECTED READINGS

Block, P., *The Empowered Manager: Positive Political Skills at Work*, Jossey-Bass, San Francisco, CA, 1990.

Johnson, B., *Polarity Management — Identifying and Managing Unsolvable Problems*, HRD Press, Amherst, MA, 1992.

Maister, D. H., *Managing the Professional Service Firm*, The Free Press, New York, NY, 1993.

Morgan, G. M. and Henrion, M., *Uncertainty — A Guide to Dealing with Uncertainty in Quantitative Risk and Policy Analysis*, Cambridge University Press, Cambridge, MA, 1992.

Pascale, R. T., *Managing on the Edge — How the Smartest Companies Use Conflict to Stay Ahead*, Simon & Schuster, New York, NY, 1990.

Welch, David A., *Decisions, Decisions: The Art of Effecive Decision Making*, Prometheus Books, Amhert, NY, 2001.

Williams, R. B., *More Than 50 Ways to Build Team Consensus*, Skylight Publishing, Andover, MA, 1993.

8 Build a Contributing Factor Diagram

THE CONTRIBUTING FACTOR DIAGRAM — GETTING STARTED

As the risk model takes shape, variables that contribute to the solution of the problem will be added. Names for these variables should be agreed upon and the names recorded on some type of movable object such as a large Post-it piece of paper or a magnetic octagon (the type used in systems-thinking exercises). These name-bearing objects should have on them sufficient room for recording of the name and some additional information. Post-it notes are relatively inexpensive and can be stuck on a vertical board for viewing and arranging. These arranged items can comprise a contributing factor diagram (CFD). An example of a CFD is shown in Figure 8.1.

The CFD of which I advocate the use is an outgrowth of the more formal and rigorously constructed influence diagram. In the classical and academically adherent influence diagram, a mix of symbols are used to represent nodes of various type. Decision nodes typically are depicted as rectangles. Ovals indicate uncertainty nodes and octagons signify value nodes. Double ovals or octagons represent deterministic nodes.

Years ago I used to endeavor to construct, strictly speaking, influence diagrams when attempting to outline the system of equations for a risk model. Almost invariably, it turned out that in the initial stages of a project I knew so little about how the various components would, in the end, interact that it was folly to attempt to deduce the esoteric linkages between the myriad puzzle pieces. Even now, after many years of experience in building such diagrams, I find I am doing really well if I can construct a rudimentary arrangement of the fundamental project components.

In a CFD, therefore, I make no veiled attempt to suppose that I have insight or foresight enough to deduce the nuances of the risk-model algorithmic relationships. The plain and unadulterated ignorance typical in the initial stages of risk-model building generally precludes the fabrication of a formal influence diagram. This leads to the use of the much simpler, for lack of a better term, contributing factor diagram.

Before influence-diagram-huggers renounce me as a risk-heretic, I would here like to adduce that influence diagrams certainly do have a prominent role in modeling. Such diagrams can be used for at least two purposes. One common application is to use the plots to outline a work process. Another application is to utilize the influence diagram to delineate the logical and mathematical relationships between risk-model variables. It is for this latter application that I recommend the simpler CFD. A detailed plant-construction example will be given later in this chapter.

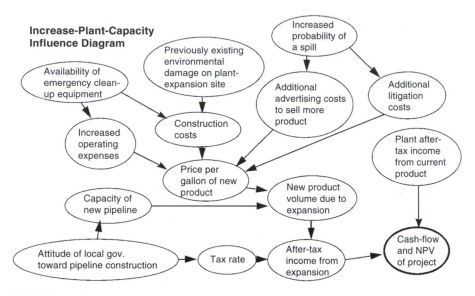

FIGURE 8.1 Example of a contributing factor diagram.

Something to realize about a CFD is that it is not a flow chart. In a flow chart, entities are arranged in sequence. Things that happen first generally are arranged at the top of the flow chart, and items that are subsequent are listed below. In contrast, the items arranged on a CFD are not time- or sequence-dependent. Items are connected with arrows that indicate that one item is used in the calculation or determination of another. For example, in Figure 8.1, the variable "Probability of Spill" is used in the calculation of advertising costs and litigation costs. The fact that these variables are posted near the top of the diagram has nothing to do with whether, in the risk model, they are used prior to or after any other variables shown.

In a CFD we generally work backward from the answer. That is, having previously succinctly defined the question and the answer variable and its units, we post the answer variable first on the CFD. Before proceeding with the posting of additional variables, however, there is one more answer-variable attribute to be discussed. The next step in CFD construction is to decide just how the answer to that question will be presented. In Figure 8.1, the answer variable is "After-Tax Income." We will express this in dollars. At the time of CFD construction, the group must come to consensus concerning just how the output variable will be displayed and presented to management. Later in this book, it will be disclosed just what is a distribution. However, it is important at this juncture to consider some practical aspects of distributions.

In Figure 8.2, the variable "After-Tax Income" is displayed by three different bar charts. The data comprising each bar chart are identical, the only difference being the number of bars we selected into which the data would be subdivided for display. It can be seen that when only five bars are selected, the bar chart has a roughly log normal appearance (skewed to the right). When 10 bars are selected, the log normal character is lost; it appears that there is a part of the distribution

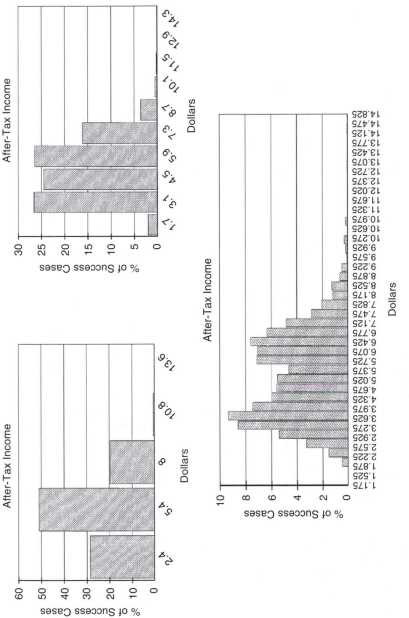

FIGURE 8.2 Frequency plots of common data using 5, 10, and 40 bins.

where the frequencies are all similar and that there are outlying values on either side of this range. When 40 bars are selected, the bar chart takes on a bimodal appearance. These three bar charts of the same data distribution certainly would convey different messages to those not well grounded in statistics.

The three different pictures, however, are not the worst of it. Consider the situation in which project economics dictate that if the project cannot generate about $5 million, then we likely will not pursue it. In the 5-bar chart, the probability of generating around $5 million is about 50% contrasted with the 10-bar probability of only around 25%. The 40-bar probability is less than 9%. It should be obvious to those versed in statistics that the differences in probabilities are related directly to bin sizes. Managers to whom the bar chart of "After-Tax Income" is presented, however, may not be so savvy in statistical inference, and certainly will not be presented with multiple plots of the same data for comparison. This topic is addressed in greater detail in Chapter 12.

At the time of CFD construction, it is essential that the group decide on just how much detail (how large or small the bins need to be) needs to be conveyed in the bar chart. Group members should come to consensus on this issue and record exactly why they decided on a level of resolution. All members of the group should agree to consistently present the "After-Tax Income" data.

Following resolution of the problem of how answers will be presented, additional variables should be added to the CFD. These additional variables should be those that contribute to the calculation of the answer variable and which have passed the tests outlined in the "Identify and Define Variables," "Ask the Right Question," and "Double-Dipping" sections of this chapter.

For each variable added, the group should agree on a written definition of the variable and just what will be asked of the user. For example, in Figure 8.1, one of the variables listed on the initial contributing factor diagram is "Attitude of local government toward pipeline construction." At first blush this may seem like the question to ask. Consider, though, that we have defined our output variable to be "After-Tax Income." Therefore, we are going to have to gather from the user information relating to local government attitude that will allow us to include this information in a calculation that results in "After-Tax Income" expressed in dollars.

Upon further consideration, we may decide that "Attitude of local government toward pipeline construction" is not exactly the question we need to pose to the user that will supply data to the risk assessment model. In this case we may want to break down the original question into the following set of questions:

- What is the probability that local government will delay pipeline construction?
- What is the likely length of time of any delay?
- What is the cost per unit time of the delay?

Users would be prompted to supply the answer to the first question as a distribution of probabilities expressed in percentage. Units for the second question related to time may be a distribution of months. A distribution of dollars per month would suffice as input for the third question. The user-supplied data then could be combined thus:

(probability of delay) x (length of delay in months) x (cost per month) (8.1)

In this way, political considerations can be used to contribute to the calculation of "After-Tax Income."

In addition to posing each consideration so that it can be used in the calculation of the identified output variable, it is essential that the group agree upon the exact wording of the question or statement that will prompt the user to enter the desired data. For example, for the first question listed above, the group might settle on this phrasing:

Enter below a distribution of decimal percentages that represents the probability that local government will delay the project due to their attitude toward pipeline construction in the project area. This probability should not include consideration that the government will cancel the project altogether, only that the local authorities will delay the project.

Wording of the question is critical. With this question, it is clear that we want probabilities of delay related only to the government's attitude toward pipeline construction in the project area. This should not include, therefore, consideration of other reasons the company might have for delaying the project (lack of funds, availability of materials, etc.). In addition, we may wish to capture separately the probability that the government will cancel the project altogether. This is a different question relative to calculating the cost of a government-imposed delay. From this simple example it should be clear that the wording of questions is critical to gathering the desired information from users.

Pertinent variables are added one at a time to an increasingly detailed and evolving CFD. As each is added, it is placed on the diagram in a position that indicates its relationship to variables already existing on the diagram. Arrows are drawn from the newly added variable to variables that use the newly added variable in their calculation or determination. A poorly designed CFD can appear like a plate of spaghetti with intersecting and crossing arrows. In a well-designed diagram it is easy to trace factors backward from the answer variable to the most important input variables. A well-designed CFD is a blueprint for generating the equations that will form the basis for the risk assessment model.

IDENTIFY AND DEFINE VARIABLES

Following the establishment of categories and the posting of the answer variable on the CFD, the next step is the identification of variables that comprise the categories and the establishment of the relationships between the variables. The risk-process facilitator should guide the group through a discussion of each of the categories to establish just what variables might be appropriate. As outlined previously, we might establish (at least) the following categories and variables for our plant-construction project:

- Economic
  - Price (dollars per gallon) of product
  - Advertising costs
  - Litigation costs
  - Construction costs
  - Tax rate
  - Operating expenses
- Environmental
  - Probability of a spill
  - Attitude of local government toward pipeline construction
  - Previously existing environmental damage
  - Availability of emergency cleanup equipment
- Technical
  - Capacity of new pipeline
  - Capacity of existing facilities

Just as categories can be broken down into sets of variables, the variables themselves can be subdivided. For example, it may be necessary to subdivide the construction cost variable into component variables such as heavy equipment rental costs, compressor (purchase) costs, and so on.

It is the job of the facilitator to ensure that any given subject or variable is not unnecessarily subdivided. Too much detail is detrimental, and the group should strive to keep variables as general and high level as possible. It also is essential that any question devised to query a user for variable-related information be posed such that it prompts the user to provide information that will best serve the purpose of the risk assessment.

ASK THE RIGHT QUESTION

Asking the right questions (in addition to correct phrasing) can be critical to success. I have, in the past, been asked to consult with groups charged with supplying clients with risk assessments for construction costs. For one of these jobs, the fundamental problem was that actual project costs deviated significantly from those predicted by the risk assessment model run prior to project inception. Analysis of the risk assessment process employed by this group revealed several problems, not the least of which was the phrasing of questions that prompted users to supply information. Some of the questions posed did not contribute to the solution of the actual problem.

The ultimate goal of the group's risk model was to predict, prior to beginning the project, the total project costs. One of the costs of such a project is that of compressor purchases. Because the risk model was intended to be run prior to the project's actual beginning, the risk model asked, among other things:

- How many compressors will be needed?
- How much does a compressor cost?

The user generally entered a single value (say, 1) for the answer to the first question and entered a distribution of costs (say, $100,000 plus or minus 10%) for the answer to the second question. These costs and calculated total costs were being generated prior to beginning the project. Therefore, projects could still be rejected because the project absolutely was too costly or because it cost more that a competing project. For these reasons, there was considerable pressure by project proponents to minimize costs. The costs for compressors are fixed by manufacturers, so there is little room for cost savings there. However, it could be proposed that a small number of compressors will be needed, thus keeping costs down.

The actual number of compressors needed for a given type of project generally increased through the life of the construction project relative to the number entered in the pre-project risk assessment model. Therefore, it should come as no surprise that the actual total cost of a project significantly exceeded initial pre-project estimates. To remedy this problem (in part), I suggested that they ask different questions with respect to compressor costs. The questions I proposed they ask are as follows:

- When projects of this type are finished, how many compressors did we typically use?
- How much does a compressor cost?

Asking these questions will yield much more accurate total-project costs, with respect to compressors, than the original set of questions. Much of a facilitator's job centers around limiting the amount of detail (yet capturing sufficient detail) demanded by the model. In addition, the facilitator should be concerned with asking the right questions. Pertinent questions should be phrased such that answers to questions result in user-supplied information that actually contributes to answering the question that is the aim of the risk assessment model.

DOUBLE-DIPPING

Generally, the facilitator is much less familiar with the problem than are the members of the group. However, a facilitator should use common sense and his or her relative ignorance to advantage. Realize that no one else in the room can ask fundamental questions with impunity. The facilitator should question the addition of each variable and consider whether the information that is to be captured by the new variable is not, in fact, already captured by some other variable or set of variables. Avoidance of double-dipping (capturing essentially the same information more than once, but perhaps in different ways) always is a major concern in model building. Often, group participants are too close to the problem to realize some of the overlaps in logic. This especially is true when representatives from different disciplines approach a common problem.

For example, it may be deemed that operating expenses are too high. An engineer may attempt to reduce operating expenses by suggesting that certain types of more expensive but more durable materials be used in the operation, thus lowering operating expenses in the long term. The economist, however, has already in his mind discounted the use of those materials because they are too costly. The economist's approach to the problem is to add a third shift, thus reducing overtime expenses. Both approaches solve a common problem (operating expenses are a bit too high). However, we would like to keep construction costs down (use cheaper materials if possible); and we would like to keep the number of employees to a minimum (two shifts are better than three). Thus, we likely would invoke one of these operating cash-reduction methods, but not both. It generally falls to the risk-process facilitator to ferret out the duplications. Getting each participant to express why he or she is suggesting a particular approach for solution to a problem and employing experience and common sense are typical ways to guard against double-dipping.

DOUBLE-DIPPING AND COUNTING THE CHICKENS

Double-dipping appears in many insidious forms. One particularly menacing embodiment of the malady is what I like to call the counting the chickens problem. This topic is covered in more detail in the Type Consensus Models section of Chapter 5, but because of its link to double-dipping, it bears mentioning here.

Double-dipping is a difficult hazard to avoid when building a model. A good facilitator attempts to ensure that no piece of information is collected, in different ways, more than once in a risk model. Avoidance of this problem is, more or less, controlled by the facilitator and the model builders. However, once the model is built and shoved out the door to the user community, double-dipping of a different type can take place.

Only organizational and process-related safeguards can even begin to avoid the counting the chickens problem. Consider that we have five houses that form a circle. They, therefore, share a circular backyard. In the yard there are a dozen chickens. We knock on the front door of each house and ask the occupants how many chickens they have. They look out the back window and report that they have 12. This same answer is given at every house, leading us to believe that 60 chickens exist. Such counting of things more than once is a common failing in the organizational implementation of a risk assessment process.

FIXING THE DOUBLE-DIPPING AND COUNTING THE CHICKENS PROBLEMS

The counting-the-chickens problem is an especially difficult dilemma to solve. There exists, to the best of my knowledge, no technical basis for its solution. Organizational awareness and aplomb tend to be the only remedies. Although a potential solution, organizational cures are not easily implemented.

Consider a company that produces a product in several locations around the world. At each location the company expects that the lion's share of the product will be consumed by the local indigenous population. The volume of product produced in excess of the amount consumed locally is expected to be exported to a targeted population of consumers. It is not uncommon to find that multiple product-production facilities have based their business plans on a common population of consumers. Those consumers might be able to absorb the excess production from one facility, but certainly not from all plants. Well, in this situation, how do we prevent the counting-the-chickens (i.e., counting the same consumers more than once) problem?

As previously stated, there is likely to be no simple or even intricate technical solution to this problem. That is, implementing a software or hardware solution probably is without merit. Organizational processes probably are the only avenues to a possible remedy, but organizational solutions can be notoriously untenable.

For example, the aforementioned product-producing corporation might decide that it will now require that all such proposals be funneled through corporate headquarters prior to actually implementing the plan. At corporate headquarters, then, it must be somebody's job to investigate each proposed production plan and to ferret out the overlaps in plans such as counting potential customers more than once.

This solution works well when the data containing the proposed plans, sent from the producing locations, are truthful and accurate. Exporting is more expensive than local consumption, so the plans may play down the amount of exports. This can lead the corporate overseer to believe that the consumers targeted to absorb exports are capable of taking up the slack for more production facilities than they actually can. In addition, the corporate overseer position is often not a popular one. It tends not to be a career-path position and is a post that is viewed with some degree of resentment by the operating facilities that are being governed. Therefore, this position typically experiences high turnover; that is, those assigned to the job generally take the first opportunity to vacate the position. Without consistency with regard to staffing of this critical assignment, much of the effectiveness of sleuthing out problem areas is lost.

Another popular remedy is to assign a person or group to become a temporary part of the production site team that is going to put forward a proposal. The assigned person generally moves to the production site and becomes part of the process and organization that will generate the proposal. The idea here is that such a person who joins multiple teams in different locations will be able to spot the problems before they become part of the proposal. This approach, however, has some obvious drawbacks.

First, the employees whose job it is to join the location-specific team must be willing to move to the remote site for an extended period of time (and repeat the process when the current assignment has ended). This can definitely be done, but it cannot be done indefinitely. In addition, more than one proposal might be fabricated at the same or different facilities. This requires that there be more than one employee assigned to such a job. If more than one person is involved, then avoiding the counting-the-chickens problem depends upon close coordination and good communication between those individuals. This, in itself, can be difficult to implement.

Although both solutions outlined here are only partial cures and are far from perfect, implementing some manner of organizational control to avoid double-dipping is better than instituting no plan at all. The counting-the-chickens problem is a particularly insidious risk-modeling problem and can wreak havoc on a company if it is not suitably addressed.

CFD-BUILDING EXAMPLE

Let's consider a situation in which Joe has to build a new chemical plant. In the initial planning stages, Joe might wish to construct an influence diagram that will depict the stages and steps that need to be taken between the present time and the event of opening of the plant. In such an influence diagram, Joe would begin by posting the answer variable — the opening of the plant. Joe knows that at least the following variables will influence the plant opening, and, therefore, need to be included in his diagram.

- Negotiations and contract signing with the labor union (negotiations may not be successful).
- Identifying local suppliers for building materials. There is no guarantee that critical supplies can be had locally.
- Governmental approvals and permits must be obtained. Permits have, in the past, been denied for similar projects.
- Engineers must submit a construction plan that is not only economical, but that will comply with environmental regulations. In addition, the plan must result in a plant that will fit on the site but will provide sufficient production capacity. This is not a given.
- Local environmental groups must be placated.

Joe will commence with his project by organizing the aforementioned items and others as an influence diagram (not shown here). Rectangles (decision nodes), ovals (uncertainty nodes), and octagons (value nodes) will be used to represent the individual items. For this type of planning, influence diagrams are invaluable.

Ultimately though, Joe will be faced with the arduous task of calculating, for example, the net present value (NPV) and perhaps the internal rate of return (IRR) of the plant. In the initial stages of the project, Joe does not have a firm grip on exactly what variables will contribute precisely to a set of equations. He does have some inkling as to the general relationships between the variables that will be used to construct the strict set of equations of the risk model. For this purpose, at this stage, a contributing factor diagram (CFD) is more suitable.

In a typical CFD, all variables are represented by a single common symbol — an oval, a rectangle, or whatever. No intrinsic meaning is attached to the symbol. Symbols are, like those in an influence diagram, connected by arcs and arrows to indicate which variables contribute to the calculation of other components (even though, at this stage, the precise equations that might be used to calculate a given variable might not yet have been decided). An initial stage of a CFD is shown in Figure 8.3.

In Figure 8.3, Joe has posted the answer variable (NPV and IRR of the new plant) and several major factors that will contribute to the calculation of NPV and IRR. Joe knows that he will have to calculate total cost and that, at least in the initial stages of the project, construction cost will be a major part of the total cost. He knows also that to have healthy NPV and IRR values, he must more than offset the costs by generating sufficient income from the plant.

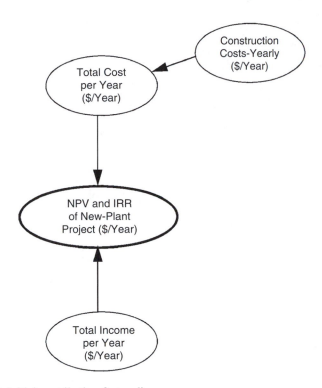

FIGURE 8.3 Initial contributing factor diagram.

In a meeting with a facilitator and all other pertinent players in the plant-construction project, Joe presents his initial CFD and asks for comments, additions, etc. One of the first things to be pointed out is that the construction cost variable needs to be subdivided into its major component parts. The facilitator, in an attempt to keep things simple, challenges the breakdown of the construction cost parameter but is quickly convinced that at least some subdivision is pertinent.

Subdivision proponents argue that if the plant project is killed in the initial stages by abnormally high construction costs, it will not be sufficient to know, and to send to a database, simply the fact that construction costs were exorbitant. It will be necessary to know whether labor or steel or concrete or whatever parameter (or combination of parameters) causes the construction cost variable to contain coefficients of unreasonable magnitude. As can be seen in Figure 8.4, the original construction cost variable will represent the combination of costs related to labor, pipe, steel, compressors, concrete, and exchangers.

Labor, it is decided, will be divided into two types. Construction labor is considered a capital expense (CAPEX) while operating labor (plant workers, office staff, etc.) is considered an operating expense (OPEX). Later in this book, in the section Comprehensive Risk Assessment Example in Chapter 17, I will use this CFD to develop the associated risk model. It will be realized at risk-model-building time that another variable that apportions CAPEX and OPEX labor costs on a year-by-year basis will need to be added as a risk-model parameter. At the current stage of

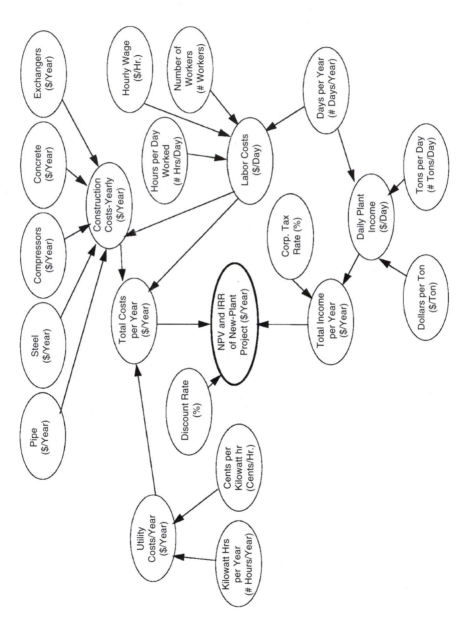

FIGURE 8.4 Completed contributing factor diagram.

development, however, Joe and his cohorts are unable to predict all the variables and relationships that will be required by the equations of the risk model — thus the CFD.

Labor, it is decided, will be calculated from these factors: days per year, hours per day, hourly wage, and number of workers. This will be done regardless of whether Joe is calculating CAPEX labor costs (the arrow between "Labor Costs" and "Construction Costs — Yearly") or calculating OPEX labor costs (the arrow between "Labor Costs" and "Total Cost per Year").

Joe's group decides that days per year not only contributes to the estimation of labor costs, but also will serve as a term in the equation that calculates daily plant income. The dollars per ton and tons per day variables also will serve as terms in the equation that calculates daily plant income.

Daily plant income and corporate tax rate will be the major contributing factors in the calculation of total income per year. Along with discount rate, this answer parameter, represented as a time-series variable, will, in turn, be the major contributing factor in calculating yearly cash flow for the NPV and IRR estimates.

In addition to labor cost OPEX and construction costs, Joe's team decides that the other major contributing factor in the calculation of total costs is utility costs per year. At the time of the initial CFD-building meeting, Joe's team decides that two factors should contribute to the calculation of utility costs per year. Kilowatt hours per year and cents per kilowatt hour are added for this purpose. However, it will be discovered later that the team decides that if utility costs are exorbitant, there is no real information, all things considered, in knowing whether the number of hours or the cents per hour is the culprit. In addition, the team, through use of the model, will discover that users are more adept at supplying total energy costs per year (as you might be if you were queried concerning your home's energy costs) and are less adept at providing accurate and precise information about hours and cents.

As you will also discover in the Comprehensive Risk Assessment Example of Chapter 17, the team will decide that the construction costs should be treated in a separate model. The output distribution from the construction costs model will be used as an input distribution for the main NPV/IRR-calculating model. The team decides this because the construction phase of the plant is imminent and the construction teams, at this stage, have to rerun the model multiple times in an attempt to emulate several construction scenarios. In addition, this cadre of construction consultants has a pretty good grip on the associated costs, but they do not have a clue concerning items such as discount rate, OPEX costs for labor, dollars per ton, and other variables. They would be obligated to supply values for such variables if the model were not divided.

One more thing to note about the CFD in Figure 8.4 is that each variable, starting with the answer variable (NPV/IRR), is labeled with units. A good facilitator will preclude the addition to the CFD (not an influence diagram) of any variable that cannot be put into a set of equations and somehow be used to calculate the answer variable. This is an important point because the risk-model builder likely will not have all team members available at the time of building the model and will have to rely on the CFD as a blueprint for model assemblage.

SHORT LIST OF HINTS FOR BUILDING A CFD

The numbered steps below are a synopsis of how to approach the building of a CFD.

1. Decide what the answer variable is to be. (This can be difficult.) There does not have to be just one answer variable, but decide on just one to start with.
2. Decide in what units you want the answer variable to be expressed. For the sake of this example, let's assume that you want to calculate NPV and that you want it expressed in dollars.
3. Begin to add large variables to the diagram that would contribute to NPV. By large I mean things like costs, incomes, etc. Then show the things that go into making up each large variable. For example, costs might consist of the two variables OPEX and CAPEX. OPEX might consist of labor costs, rental fees, etc. CAPEX might be composed of the cost of steel, the cost of land, etc. Likewise, incomes might include number of barrels sold, price per barrel, etc.
4. For each variable that you add, be sure to decide in what units it would be expressed. For example, if you add a variable to the diagram called politics because you think there might be political problems with the project, consider that your output variable (NPV) is expressed in dollars. How would you include politics in an equation that would result in dollars? Well, I would change the politics variable to three new variables. They would be as follows:
   a. The probability (percent chance) that we will have a political problem (expressed as a distribution of values between 0 and 1).
   b. The length of time we might be delayed if we have a political problem (expressed as a distribution of days).
   c. The dollars per day it will cost us (expressed as a distribution of dollars) if delayed.

Now I can multiply the percent probability of delay times the length of delay times the dollars per day and get a dollar value that can be used in an equation to calculate NPV.

Likewise, do not put a variable on the contributing factor diagram that says, for example, electricity. What you really want is kilowatt hours expressed as a distribution of hours and cents per kilowatt hour expressed as a distribution of cents. These can be combined to influence NPV (in dollars). If you use the logic outlined above, you should end up with a contributing factor diagram on which each variable can contribute to an equation or set of equations that will, eventually, calculate cash flows and then NPV.

SELECTED READINGS

Power, D. J., *Decision Support Systems*, Quorum Books, New York, NY, 2002.
Welch, David A., *Decisions, Decisions: The Art of Effective Decision Making*, Prometheus
    Books, Amherst, NY, 2001.

9 Education — Critical at Three Levels

EDUCATION FOR MANAGEMENT

Without exception, the most critical element in risk-process implementation is that of education. As will be delineated in the following paragraphs, education at (at least) three levels is necessary. The three levels are presented separately in three sections of this chapter.

There is not much point in training the troops to think stochastically and to generate sophisticated risk models if the decision makers (usually management) are ignorant with respect to the nuances of and output from the risk process. A risk-savvy employee is not doing anyone any favors by plopping down on the unsuspecting manager's desk a menagerie of probabilistic plots, risk-weighted values, and other not-so-easy-to-understand stuff. In any organization, if the decision makers (1) are not prepared to deal with the input and output from a risk assessment; (2) do not know what questions to ask; and (3) do not know what to look for and what to look out for, risk-wise employees can expect embarrassed and put-upon management to receive coldly any foreign-looking risk-related information.

Employees should not expect unprepared managers to embrace their stochastic overtures warmly. In addition, the purveyors of the risk process should not expect managers to sit still for multiple-day classes and workshops in which are conveyed the many details of risk-model building. Education of managers with respect to risk assessment is a special process that must be handled deftly and with some organizational and psychological aplomb.

I have learned the hard way that any course designed for managers must be short in duration — usually no more than 1 day. That is not to say that special department-specific seminars in retreat situations should be discouraged or discounted. However, if a large number of managers is to be trained in the ways of risk, a course must be designed that can be inserted into a busy schedule with minimal interruption. Usually 1 day can be set aside. More than 1 day becomes a problem.

It also is important that risk education for managers be able to address any number of attendees. This is practical because there are occasions when a very small group of managers (sometimes just one) from a specific organization will request such training. Therefore, any class that involves role playing or interaction between class attendees will likely be less applicable. Any exercises offered in the class should be able to be performed individually. Attendees should not be required to have out-of-the-ordinary computer or other skills.

It should be pointed out at the onset of the class that the risk-for-managers education is not a dumbed-down version of a regular risk class. This should not only be said, but it should be true. In my classes for managers I always present the analogy of airplanes or ships and the people associated with them. There are those who build planes and ships. Those people need to understand all of the details about how an aircraft or ship is constructed. However, it is unlikely that you would want any of the builders at the controls on a stormy night. These plane and ship builders are equivalent to those employees who attend the regular risk classes in which are conveyed the concepts and practices of risk-model building.

On the other hand, I am not sure you would like the captain of the plane to attach the wings. His role is different. True, such people have to have sufficient knowledge of construction so they do not make unrealistic demands of the craft, but their primary purpose is to guide the plane or ship safely to its destination. So it is with managers and risk assessment. Most managers will not be able to write the computer code necessary to construct a risk model. However, they should be comfortable with risk-model output, they should know what questions to ask, they should share a common jargon with the risk-model presenters, and they should be capable of making decisions based on the output from multiple risk models. Therefore, it needs to be stressed that the risk class for managers is not a dumbed-down version, but rather is a class with a completely different slant than the classes offered to those who will learn to build risk models.

Nothing sells like success. An education course for managers should invariably commence, after introductory material, with a parade of examples from across the organizational spectrum illustrating where and how risk technology has been applied and the benefit it has provided to the organization. At least two major benefits result from such a show and tell. First, class attendees will realize that people not unlike themselves have put this process to use to great advantage. Reduction of apprehension is key. Second, some of the class participants will have come to the class wanting to know about risk assessment, but really have no idea how they might use it in their organization. A broad-based demonstration of what others have accomplished should be designed to allow them to see themselves as potential participants in the risk initiative.

Because a class for managers likely will be given often and because the potential audiences will represent quite diverse entities, it is impractical to tune more than just a small part of the class directly to address the specific problems of any given group. Therefore, examples shown and exercises performed by the class members should be as generic as possible. Lecture and exercises should be geared toward imparting the fundamentals of the risk process, such as the following:

- What is risk?
- How does it relate to decision making?
- What is uncertainty?
- What is a contributing factor diagram?
- What is a distribution?
- What is Monte Carlo analysis?
- What is a frequency plot?
- What is a cumulative frequency plot?
- What is chance of failure and why is it important?

- What is dependence and why is it important?
- What are risk-weighted values and what can I do with them?
- What is sensitivity analysis and why is it important?

It is important in such education to avoid getting sidetracked into discussions of how most of these risk-related items work. For example, a discussion of how a distribution is built or how a Spearman rank correlation is calculated should not take up class time. Emphasis should be put on what these things are, why they are important, and what managers can do with them.

At the conclusion of the class, managers should be given a short list of items that they can pass along to those who might present risk assessments to them. It is suggested that this list be given to all potential presenters well in advance of any risk assessment presentation and that the presenters be prepared to discuss each of the items on the list. This is a powerful list and process. Even if the manager has by the time of the presentation forgotten some of the risk-related details, presenters will not show up unprepared. They will have done their homework with respect to generating a credible risk analysis. The list also aids in consistency. With the list, everyone knows just what is expected, and similar information is presented by diverse groups for a wide range of projects. This greatly aids the manager in the process of comparing and ranking projects of similar or different types. A list might look like the one shown below.

When reviewing risk assessments, demand the following:

- To have described just what is the question being answered by the risk assessment — demand specificity
- Consistency in evaluation and a consensus model if multiple opportunities are being compared
- To see a contributing factor diagram
- A discussion of chance of failure and how it was handled
- A discussion of dependencies between variables (in the contributing factor diagram)
- A discussion of risk-weighted numbers (if appropriate)

This is all part of the *education* of decision makers — which is distinct from *training*. This topic will be addressed in detail in a later section of this chapter.

INTRODUCTORY CLASS FOR
RISK-MODEL BUILDERS

A second level of education and, thus, a second set of classes has to be designed for those who want to know how to build a risk model, but presently know nothing about it. Like the class for managers, these classes should use examples that are as generic as possible so that engineers, economists, scientists, and people representing other disciplines can relate to the lecture and exercises. Unlike the manager's class, this class should be up to 2 days in length.

Classes aimed at the novice risk modeler need to begin with the very basics. However, unlike the manager's class, care should be taken to delineate, to a practical degree and depth, just how things actually work. For example, just how does the software package you are using generate each type of distribution? Exactly how do you implement dependence, and what equations are used to do so? What are the equations behind the calculation of the various types of risk-weighted values? Through example, lecture, and hands-on exercises (this is the *training* part), class attendees should gain reasonable understanding and practical experience related to all fundamental aspects of risk assessment.

Individual lectures and exercises should be designed to present each salient aspect of risk assessment beginning with the decision-making and contributing-factor-diagram processes. Terminal exercises should guide participants through the construction of meaningful financial measures such as net present value. A suggested plan of attack for a class might be as follows:

- Discussion of decision-analysis/making tools such as contributing factor diagrams, sorting processes, and other applications
- Definition of risk and uncertainty
- Discussion of how risk technology is being used in this business and in other businesses
- Distributions — all about them
- Monte Carlo analysis
- Frequency and cumulative frequency plots
- Chance of failure
- Dependence
- Time-series analysis
- Sensitivity analysis
- Risk-weighted values

Upon completion of the course, attendees may not have the skills to build sophisticated risk models alone — they likely will still need help with that. However, class attendance should produce a crop of users who are no longer intimidated by the concept of risk assessment, who understand the fundamentals of risk technology, who feel comfortable with whatever software is being utilized, and who can construct basic risk assessment models. If your class accomplishes this much, you did good.

ADVANCED TRAINING FOR RISK-MODEL BUILDERS

A third level of training has to be aimed at the relatively risk-astute people in the organization. Those individuals who have taken the introductory course and those who have some previous background in statistics or risk assessment require a special class. For those who have taken the introductory class, this advanced training should reinforce concepts and challenge participants to put to work what they have learned. For those who have not been previously indoctrinated in a company class, this course should contain enough review material to introduce these people to how we do things, but not so much review as to bore to tears those folks who have attended the introductory class.

The format for the advanced class should minimize lecture and maximize exercises and problem solving. A format I have found to be successful is to provide the class members with increasingly difficult problems to solve. After each person has individually solved the problem, the instructor should review how he solved it and then solicit alternate solutions from the class. Lively and instructive discussions often ensue.

Initial problems might focus on simply having class participants interpret what types of distributions might be used to represent a host of variables in the presented problem. Subsequent problems should introduce complexities such as dependence, chance of failure, and other risk-related techniques. Final problems should include time-series analysis and the calculation of financial measures such as net present value, internal rate of return, and others. Classes of this type concentrate mainly on *training* as opposed to *education*.

ORGANIZATIONAL ATTITUDE AND TRAINING CENTERS

A complete understanding of the education and training facilities and policies within the organization also is important. This aspect of risk-technology transfer often is not considered in the design of a new process. No new process should be generated without a complete understanding on the part of the technology generator as to how people will be trained to use the product. It is a mistake to assume that potential users will want to use the product or, even if they want to, that they will be able to successfully use the new technology. Training is absolutely vital.

Consideration of the training aspect should come early in the product design. It is not enough that the new process be easy to use. A fundamental error committed by many technology generators is they assume that because the new process is easy to execute, users somehow will have the same fundamental understanding of all aspects of the process as is the risk-technology creator. Nothing could be farther from the truth. The only safe way to field a new process is to make sure that the users fully understand all aspects of the new risk technology. This can only be done through training classes. Never assume that users will read documentation — *tell* them and *show* them.

Introduction of risk technology often requires the concomitant introduction of new hardware or software. It is worth mentioning here the relationship between requiring new equipment for a product and a training program. Training within organizations is accomplished by a wide variety of processes and mechanisms. To implement a successful training program, you must have a good understanding of how things work with respect to training facilities and policies. No matter what the process, it is likely that until the product has proven itself in financial terms, the person introducing the risk technology and associated management should expect to have to supply the required new equipment for the training program. If training takes place at the risk product creation site, then this likely is not a major problem (although it is more expensive to send multiple trainees to a single site than it is to send a single trainer to multiple sites). Likewise, if receiving site management has agreed to purchase the required equipment, then training at that site also should not be a problem.

Many organizations handle training through central training centers. Training centers can be financially autonomous, and it may be difficult to convince them that major funds should be spent to provide training for a new and unproven process. In the case of the resistant training center, there are two practical solutions to this problem. Either the risk product creation site has to provide the required equipment, or the product-receiving site has to be convinced that it is in its best interest to fund the purchase for the training center. Receiving sites may buy unique equipment for the new risk technology because they are convinced of the value of the new process, but it will be a rare instance when a training group or facility will make the same accommodation. These practical problems should be considered early in the risk technology transfer process.

Some organizations have no training policies at all. If the organization has an established training procedure, it is critical that the risk process creator insist upon doing the training or (if he or she is, for example, a terrible speaker) that the process creator control the course content and who teaches the class. It also is most practical if the process creator can design the training classes so that they can be taught in available facilities and on available hardware. Training should not be the last thing to be contemplated. When considered at the onset of risk technology development, training procedures can be woven into the design of the process. Training is essential to the process of risk technology transfer.

TIME DEMANDS AND ORGANIZATION OF THE EDUCATION EFFORT

In life in general, the time it takes you to complete your tasks can depend more on your degree of organization than on the absolute amount of work to be done. So it is with education and training associated with launching a risk assessment effort. One of the things that shocks risk process proponents most is the amount of time that actually needs to be devoted to the educational aspect.

Astute organization of the educational facet can help reduce the time investment. However, even the most fastidiously organized individuals will attest to the fact that the educational process is eating their lunch. Shortcuts in the organizational effort can lessen the impact of risk education.

For example, in my line of work I have the pleasure to interact with individuals and groups that represent a broad spectrum of disciplines. Within a short time span, it is not unusual to meet with and teach classes to lawyers, accountants, managers, economists, manufacturing personnel, engineers, scientists, and others. If it becomes your charge to interact with such a large number of diverse disciplines, there are basically two ways you can go with regard to generating training materials or examples.

The first method is to prepare classes that contain detailed risk-related exercises that address specific areas. Using this approach, you present general generic material in each class but design hands-on exercises that are discipline-specific.

One advantage to this approach is, of course, that the attendees can relate directly to the material and, therefore, absorb more of what is taught. In addition, class members walk away with practical go by examples that they can put to use as soon as they return to their jobs. This approach tends to increase the popularity and demand for the class within each discipline due to favorable word-of-mouth reviews.

It should be clear that the disadvantages of this method are mainly yours. Attempting to generate meaningful (and not goofy) complex, risk-related examples for a multitude of disciplines (or even one if you know nothing about it) can be time-consuming and discouraging. This approach also puts the onus on the class developer to keep up with changes and advances in each discipline. Things change rapidly, and it does not take long for a hard-earned set of examples to become passé. Reworking a class can be more time consuming than setting up the class in the first place.

I have known others who have taken the area-specific example route and hired experts in the various areas. Experts are brought on board to design or help design meaningful class examples. This approach relieves the class presenter of the burden of creating exercises of substance, but usually leaves them with the task of presenting the material. Not many other things you can do will more rapidly deflate your credibility with the class than presenting material with which you are obviously uncomfortable. For example, a room full of accountants will perceive in one second that you really do not understand the material that you are presenting (material that was prepared for you by another accountant). You are the teacher. If you do not understand a subject better than those to whom you are presenting, and if your knowledge of a subject is a mile wide and an inch deep, it is far better to avoid that situation entirely.

Classes with discipline-specific exercises also can be co-taught by the risk expert and by an expert in a given field. This approach works well when the expert in the field also has a grip on the fundamentals of risk technology. If the expert does not, then his presentation, in the case of the accountants, will tend to be simply an accounting exercise plopped into the middle of a risk class. Experts also should be presenters of quality and should be available when you need them. Good luck!

I have chosen to take the course of using the same set, with minor exceptions, of generic and general examples for every class. In each class, I incorporate a mixture of general business, political, and environmental exercises with which I am comfortable and about which I am knowledgeable.

The advantages of this approach are many. First, this practice limits preparation time — each class of a given type draws on the same examples no matter the makeup of the audience. Nobody in the room knows more about the examples than I do (or at least I like to cling to that belief). Because I composed and understand the examples, they can, with a minimum of effort and without having to seek counsel, be updated and enhanced. When presenting material with which I am familiar and comfortable, I can concentrate on a relaxed, humorous, and interactive delivery. I neither fear questions nor am I skittish about delving into the details of a situation.

Another real and practical advantage to this teaching method is that when on the road traveling from one teaching engagement to another, using a common set of examples greatly reduces the volume of physical stuff I have to haul with me. You will not be able to appreciate this point until you are the subject of those disapproving looks from airline employees who are eyeing your carry-on dunnage.

Using common sets of examples also has its drawbacks. If, for example, the audience is composed of a room full of accountants and none of the class examples deal specifically with accounting, taking this approach counts on each class member's ability to translate the class work to his own situation. I attempt to mitigate this foible by asking of the class, after each example, if there is anyone in the audience who can see a practical and immediate application of the principles just presented to what they do. Such a query never fails to produce a crop of raised hands. Turning the floor over to several class members to relate their viewpoints almost always makes clear the link between the exercise and the discipline represented by the class. An advantage to this approach is that I do not have to have a clue concerning what they are talking about.

Another related drawback is that I am sure that class attendees do not absorb as much of the doled-out information as they would if the class patter were couched in their jargon. Some things you just have to learn to live with.

With regard to examples and exercises, the generic versus class-specific approach represents one proverbial fork-in-the-road dilemma in which you have to choose which way to go. Another such bifurcation is the choice between regularly scheduled mass training sessions and a call me and I will come to you approach. Certainly in every training effort there is a mix of both methods, but one must dominate.

Using the mass training doctrine, the provider of classes generally arranges with training centers in various locations to hold classes on specific days. This schedule, then, is announced to the world, and those interested in the subject are encouraged to attend.

This approach to teaching has several salient benefits. First, people can see a schedule of classes that extends relatively far into the future (usually a year or more). This allows them to plan ahead and work the classes into their schedules. This also is true for the class instructor. Regularly scheduled classes also help prevent running around like a nut on the part of the teacher, trying to bring classes to organizational cul-de-sacs populated by small numbers of bodies.

Picking up on the immediately preceding point, the mass training approach is more efficient with respect to being able to reach the greatest number of people with the least effort. It is not, however, the most efficient in other respects that I will elucidate in later paragraphs of this section. A typical call me and I will come to you class will have in it no more than 5 to 20 persons. Properly promoted mass training classes usually have a student count of 100 or more. The presenter's travel is reduced and the impact per trip is greatly enhanced.

Still another advantage of the mass training method is consistency of the message. If the aim of a training effort is primarily to bring a common jargon and understanding of the subject to an entire organization, then presenting the same material a minimum number of times aids in achieving this goal. A message repeated a great number of times to relatively small numbers of people tends to wander, despite the best efforts of the presenter to be consistent.

The mass training approach, however, also has its minus column. With respect to total cost, it is likely not the most efficient. Although under the call me and I will come to you scheme the presenter makes many more trips, those relatively paltry costs are greatly exceeded by the expenses related to having 100 or more people travel to a common site for a multiple-day class. In addition, classes taught at the customer's location can usually be held in a room or facility that costs very little to occupy for a day or two (typically a conference room). The mass training approach invariably requires bearing the expense of space rental (not to mention computers, projectors, etc.). Mass training also robs the presenter of the opportunity to experience how various parts of the organization operate. Personal one-on-one interaction is sacrificed.

Another drawback is promotion. In a previous paragraph I alluded to a properly promoted mass training class. Advertisement of such classes can be exceedingly difficult to facilitate. Even with email lists and modern communications, making everyone in an organization aware of the class content and class schedule is a very tough thing to do.

The advantages of the call me and I will come to you approach are fundamentally the opposite of the disadvantages associated with mass training. This method is cheaper to enact organization-wide. It promotes personal contact and gives the class presenter the opportunity to visit and learn about the various arms of the organization. On the other hand, this approach to training is hard on the presenter with respect to travel. Over the same amount of time, this teaching method will reach far fewer people than will the mass training approach.

Through the years I have developed a teaching philosophy that incorporates both approaches. I attempt each year to schedule as many mass training sessions as is practical. When someone or some group calls to arrange a class, I make every effort to divert them to the mass training class nearest in time and space. If, for whatever reason, they claim attendance is not possible, I then weigh other schedule demands against their needs and make a call. Sometimes I say no. Everyone in the organizational education game will have to define for himself some satisfying scheme.

Another point on education. Try not to be amazed at the time and effort required. Generating a lively and cogent class from a blank piece of paper is hard enough. (You can say a lot of stuff in an hour.) However, the biggest drain is the travel and time it takes to carry out an educational assault. Beware when you start this thing that you are creating a monster. Plan accordingly.

DIFFERENCE BETWEEN EDUCATION AND TRAINING

In the sections above, I have alluded to both education and training. There is a difference. With regard to practitioners — those who will actually implement risk-assessment processes, run software, etc. — both education and training are essential. Managers and decision makers, however, should not be, in general, subjected to the training aspect.

Education is all about why risk assessment and management are important, what they can do for your business, how such processes can be integrated into the existing business method, and the like. Managers and decision makers need to be convinced — through the education process — that the time, money, and effort required to implement a risk-based process is worth the expenditure of resources. As mentioned previously, the education process should not only appeal to their common sense, but should be littered with testimonials and examples of how implementing such a methodology actually positively impacted the decision-making process.

The appeal to common sense should focus not on the potential monetary benefits associated with adoption of the risk-based method, but should concentrate on how such a method positively impacts the business process. It is difficult to argue that the inherent benefits of implementing a risk-based method are not beneficial. A short list of these benefits might be as follows:

- Early identification of risks
- Identification of uncertainties and ranges of probability and impact
- A holistic approach which integrates all aspects of the problem
- Early setting of mitigation plans and accountabilities
- Realizing the range of possible outcomes
- Realizing where management-set target values fall in the range of possible outcomes

All of these things and more should be paramount aspects in examples of successful implementation in real-life businesses. If managers and decision makers are convinced that the methodology is sound and will in fact benefit their business, transference of the risk-based process to the new business will be significantly enhanced.

As previously mentioned, training should be reserved for practitioners. The training aspect includes practice of facilitation skills, outlining the problem, risk-identification processes, hands-on design and building of risk models, and other critical aspects of the process — all of which have been outlined and detailed in previous sections of this book. Training of decision makers is detrimental to the overall effort as is only educating those who will actually have to implement the proposed process. Be sure to make a concerted effort to separate the educational and training aspects and to identify those individuals who would best fit into each category.

SELECTED READINGS

Alan, M., *The Corporate University Handbook*, American Management Association, New York, NY, 2002.

Pascale, R. T., *Managing on the Edge — How the Smartest Companies Use Conflict to Stay Ahead*, Simon & Schuster, New York, NY, 1990.

Pasmore, W. A., *Creating Strategic Change — Designing the Flexible, High-Performing Organization*, John Wiley & Sons, New York, NY, 1994.

Piskurich, G. M., Beckschi, P., and Hall, B., *The ASTD Handbook of Training Design and Delivery, 2nd Edition*, McGraw-Hill, New York, NY, 1999.

Senge, P. M., *The Fifth Discipline — The Art and Practice of the Learning Organization*, Doubleday Currency, New York, NY, 1990.

Van Adelsberg, D. and Trolley, E. A., *Running Training Like a Business*, Berrett-Koehler Publishers, San Francisco, CA, 1999.

Wheatley, M. J., *Leadership and the New Science — Learning about Organization from an Orderly Universe*, Berrett-Koehler, San Francisco, CA, 1993.

10 Risk Assessment Technologies

DIFFERENT APPROACHES

Risk assessment models come in a variety of types. Fate-transport, political, technical, financial, and other varieties of models can be constructed. Regardless of the application, risk can be evaluated by applying a wide range of technologies. Risk assessment itself can loosely be defined as the application of any technique that deals with probability in some technically sound manner. Some of the more popular risk assessment techniques applied are Bayesian analysis, discriminant-function analysis, factor analysis, neural nets, Monte Carlo analysis, and others. Although this book will concentrate primarily on Monte Carlo applications, I will here say a few words about some other popular technologies.

DISCRIMINANT FUNCTION ANALYSIS

Discriminant function analysis (DFA) has many uses, but the primary application generally is to define boundaries (equations) that separate predefined populations of objects in n-dimensional space. DFA also defines (quantifies) the degree (probability) to which the objects belong to the populations. The defined discriminant function(s) can be used to then determine for a new object the population to which it belongs and the associated probability (the degree to which it belongs to the population). Because DFA has no ability to separate objects into groups, it requires objects (data points) that already are assigned to groups and requires measurements on variables associated with each of the objects. An example might be a group of animals that have been subdivided into nocturnal and nonnocturnal types. On each of these animals, we may have measured height, weight, sex, diet, skin covering, and so on (variables).

Although understood since the publication of R. A. Fisher's work in the thirties, DFA gained popularity as a probabilistic technique only after the advent of high-speed computers. As stated above, DFA typically is used to define the boundaries between populations in n-dimensional space and, using one of several distance measures, determine the degree to which each member of a given population belongs to that population.

To illustrate the space-partitioning functionality of DFA, we will use a conceptual model. Consider two sacks of potatoes hanging side by side in a doorframe. We will designate the sacks as sack A and sack B, each containing 50 potatoes. Each potato represents a data point. If we project the position of each potato onto either the right or left parts of the frame, because the sacks are hanging side by side, the projections of the two potato populations completely overlap. Even if we lower or raise one of the sacks half a sack length, the projections of the potato positions on the vertical part of the doorframe still will overlap. In the area of overlap, we cannot be sure if the projection of a particular point is a projection from sack A or B. If, however, we project the potato positions either on the floor or on the top of the doorframe, the A and B populations can clearly be distinguished.

It should be evident that if our sacks took on complex shapes or were oriented in the doorframe in a configuration other than side by side, we would have to find another plane in n-space (other than the floor or the top of the doorframe) that, when projected upon, would clearly separate the two populations. This is the task of discriminant function analysis. DFA attempts to find a function in n-space that, when used as a backdrop for data-point projection, will do the best job possible of maximizing the distinction between the populations. As you might imagine, as data-set shapes become complex and as the data sets between which the discriminant function must distinguish become greater in number, the complexity of the discriminant function increases dramatically.

One of the primary objectives of DFA is to examine the distribution of data in n-dimensional space (in this simple conceptual example, 2-D space) and establish a new axis upon which the data points can be projected to optimize distinction of the multiple groups. In our potato-projection example, projection of the two populations on the floor clearly separates the groups. In an actual DFA, the floor would represent the discriminant function. This function is an equation that can, in this case, use the x- and y-position values for an observation (potato) to determine to which population the object belongs.

Using a distance-to-population-center measurement in n-space, DFA also will calculate the probability that (or the degree to which) the observation belongs to the assigned population. This probability can essentially be thought of as a measure of how far the object plots in n-space from the gravimetric center of the population. DFA calls this probability a posterior probability. In the case of our doorframe, the discriminant equation would be a simple equation for a line. In most DFAs, equations are complex and represent convolute surfaces in n-space that separate populations.

Posterior probability values are risk assessment tools. For example, if a discriminant function is established for our potato problem and then a 101st potato is evaluated by the function, DFA might assign the new observation to the sack A population with a posterior probability value of 0.2. Posterior probabilities generally range from 0 to 1.0, respectively indicating a weak and strong confidence of belonging to the assigned population. In this example, the new potato was assigned to the sack A population. The 0.2 posterior probability (20% probability) indicates a relatively weak relationship between the newly evaluated potato and the majority of potatoes that constitute the sack A population (the new potato plots relatively far from the gravimetric center of the sack A population). In this and other ways, DFA can be used as a powerful risk assessment tool to help determine the probabilities associated with a wide range of events.

BAYESIAN ANALYSIS

Bayesian analysis facilitates the use of new information to update (modify) initial probability estimates. It also can use historical probabilities to revise the probability estimates associated with a new project or process. It is a powerful risk assessment and management tool. Bayesian analysis generally requires that each component of a project or process have an associated (estimated, at least) probability (chance of happening).

Bayes' theorem and associated equations were devised by the 18th-century English clergyman and philosopher Thomas Bayes as part of his study of inductive reasoning and logic. It is primarily used to analyze the probabilities associated with the variables that comprise any process or project.

Most projects and processes conclude in a result. When the project or process is completed, the final result is obvious. This final state, however, generally is not so obvious at the onset of the work. That is, at the beginning of the project, there are usually several potential or probable outcomes (i.e., a project may succeed or fail, the introduction of a new product may result in a wide range of possible sales scenarios, etc.).

At the beginning of a project or process, the initiators of the work may assign probabilities to each step of the work and a total probability can be calculated for each of the many possible outcomes. A simple example might be the environmental remediation of a drilling site. Prior to the remediation effort, there are many variables to consider which constitute the remediation task. Some of these might be the issuing of permits by state regulators, the cost of sample collecting, the cost of sample analyses, the purchase of special equipment if the analyses show abnormally high concentrations of toxins, and other factors. Each of the states of these variables may be assigned a probability. Probabilities assigned before the actual start of a process or project are called prior probabilities.

For business reasons, the company responsible for the remediation may be interested in the total probability of two states. These might be whether or not the cost of the remediation is going to be more or less than $250,000. To determine the probability of the remediation costing more or less than $250K, the company can assign dollar values and probabilities to each possible state of the variables. For example, the total cost of the project without considering sampling costs is $200,000. It may be determined that there is a 60% chance (probability = 0.6) that the state will not issue the permits, given the present remediation proposal. If the state does not issue the permits, it will cost an extra $25,000 to modify the proposal. Based on experience with the selected contractor, there is an 80% chance (0.8 probability) that the sampling will cost about $40,000. Similarly, probabilities and costs can be assigned to all other pertinent variables. We can calculate all relevant combinations of probabilities and dollar values to arrive at a probable cost. Figure 10.1 shows a graphical example of this process.

Bayes' theorem also allows us to modify or update the prior probabilities as we learn new information during the course of a project or process. The actual equations associated with Bayes' theorem can be found in nearly any statistics book and will not be addressed here. Suffice it to say that the equations simply facilitate the integration of new information (new probabilities) that modifies the prior probabilities associated with the variables and with the various final-outcome states. In the example given here concerning a drilling site remediation, the company may learn that new legislation is pending in the state legislature that has an associated proba-

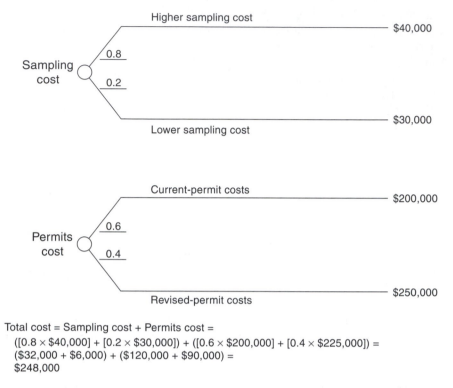

Total cost = Sampling cost + Permits cost =
    ([0.8 × $40,000] + [0.2 × $30,000]) + ([0.6 × $200,000] + [0.4 × $225,000]) =
    ($32,000 + $6,000) + ($120,000 + $90,000) =
    $248,000

FIGURE 10.1 Example of combining probabilities and consequences.

bility of significantly altering the permit issuance process. Bayes' equations facilitate
the introduction of the new information and the updating of the probabilities associated with the various end states. This is a powerful business and planning tool.

An associated use of Bayes' work is to update the probability attached to a new
object based on the prior probabilities associated with similar historical objects for
which the outcomes have been determined. An example can be found in the pre-
drill estimate of the total chance of success associated with oil and gas prospects.

A given company may be employing a prospect-evaluation risk model that is
composed of nine variables. Each variable may have an associated chance of failure
(see Chapter 13 for a full treatment of the chance-of-failure concept). The nine
chance-of-failure values are combined to give a total chance of success for a prospect.
The company presently is using the nine-parameter risk model to evaluate a new
prospect in the North Sea for which they have estimated chances of failure for each
of the nine variables and, in turn, a total chance of success for the prospect. However,
they now wish to use Bayesian analysis and historical data to validate the prospect's
calculated chance of success.

In the company's database they have analogous North Sea wells that have been
drilled. These drilled wells represent both successful wells and dry holes. A major
subset of these wells will be those for which their pre-drill estimates of the nine
chances of failure and the total chance of success were representative of what was

the eventual outcome for each well (i.e., their pre-drill estimates were on target). Baysian analysis provides them a means to use this subset of historical and analogous prospect pre-drill information to update or modify the new well's chance-of-failure estimates and the new well's total chance of success. This is a powerful risk assessment tool that can use analogies to predetermine the probabilities associated with a new project.

A third, but by no means final, application of the work of Bayes is related to decision tree analysis. This application of Bayesian-like analysis facilitates the classification and characterization of objects upon which we have measured the values of several variables.

Let's consider a sludge pit example. On five pits we may have measured pit area, the pH of the fluid in the pit, and the percent clay in the material lining the pit bottom. Two of the five pits have developed leaks. A decision-tree-like plot can be constructed to characterize (hopefully) those pits that leaked and those that did not. Such a diagram is shown in Figure 10.2.

In Figure 10.2, leaking and nonleaking pits have been segregated and characterized. Probabilities could be substituted for the numerical values in Figure 10.2 so that Baysian methods could be used to calculate the probabilities associated with each of the end states. Diagrams (decision trees) like these also can be used to determine, in this case, whether a new sixth pit is likely to leak or not.

DECISION TREES

In the immediately preceding section I defined just how a fundamental decision tree is constructed and illuminated the salient qualities of decision trees in general. I will, in the immediately following paragraphs, describe some inherent limitations in their application.

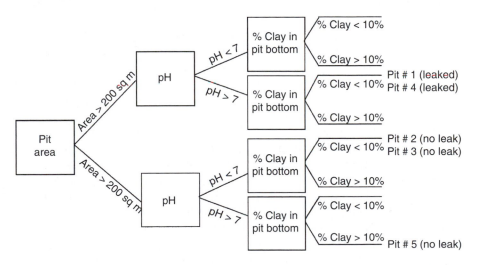

FIGURE 10.2 Decision tree for the purpose of discriminating between pits that have leaked and those that have not.

Decision trees are the near-perfect vehicle for graphically describing a series of decisions that will result in a course of action. However, decision trees are not a panacea when applied to analytical processes related to the same problem. For example, we may be charged with the responsibility for deciding whether to build a new plant (or factory). Our initial attempt to arrive at a decision will be a rudimentary one including only the following variables:

- Customer base (CB) in units of MM people
- Percent of customer base captured (PC) in units of percentage
- Operating costs (OC) in units of $MM

In our decision tree we will consider the variables in the order in which they appear in the list above. Therefore, we will first consider whether there is a sufficient customer base. Demographic studies indicate that there is a 90% chance that the customer base is of sufficient magnitude. Similar studies by consultants have indicated that there is an 80% chance that our company will capture enough of the customer base to make a reasonable profit.

If our customer base calculations have favorable results, we will consider building a plant. In our admittedly simple example here, we will consider only one more variable in our decision tree. (When we consider the analytical-type decision tree in the following text, it will be clear why we are here limiting this study to only three components.) The third variable will be operating costs.

Customer demand for our product and the announced plans of our competitors are causing us to consider building the plant sooner than we would like. At our existing facilities that manufacture our product, we employ some relatively old and somewhat outmoded technology. Our research department has informed us that new, cheaper-to-operate technology is currently being tested, but may not be ready for production-scale use by the time we have to decide on the type of technology we will install at the new plant.

If we install our traditional manufacturing process, we estimate that our operating costs will be prohibitively high relative to our competition, who no doubt will employ the latest technology. We might have to utilize older and more expensive methods at the new plant in order to beat the competition to the marketplace. Therefore, our experts estimate that we likely have about a 60% chance that operating costs at the new plant will be prohibitively high.

I consider there to be two basic types of decision trees (others may disagree — that is their prerogative) that commonly are built. These I will term the conceptual tree and the analytical tree. To construct a conceptual tree, only probabilities and single deterministic leaf-node (the extreme right-hand end of any branch) values are required. A typical conceptual tree for the problem outlined above appears in Figure 10.3.

In Figure 10.3, the aforementioned probabilities are indicated on the appropriate branch. So, in the case of customer base, there is a 90% chance that the customer base will be sufficient to sustain our business. Conversely, there exists a 10% chance that the base will be too lean and our enterprise will fail.

Considering the percentage of the customer base we are likely to capture, our consultants have estimated that there exists an 80% chance that, if we move quickly,

CB = Customer Base
PC = Percent Customer Base captured
OC = Operating Costs
STOP = Failure of project due to unacceptable values for one of the components
−40 = Estimated yearly profit from project
10 = Probability of a given branch

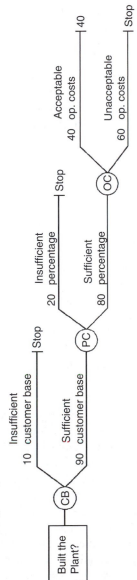

FIGURE 10.3 Example of a conceptual decision tree.

we will capture a sufficient share of customers to make our new plant a successful economic venture. The 20% chance of capturing an insufficient number of customers reflects the facts that the customers, for whatever reason, might largely shun our product and/or the probability that the competition might beat us to the marketplace. As already explained, we believe there is a significant chance (60%) that our operating costs will exceed our acceptable limits. This is duly noted on the decision tree.

At the far right-hand end of the tree is a leaf node to which a consequence has been assigned. In this case, the consequence value is the deterministic amount of money (in millions of dollars) we expect a successful plant to generate.

Back at corporate headquarters, it has been decided that, to be economically feasible, the plant must generate at least $10MM in income per year. To determine whether or not this enterprise is capable of infusing the corporate coffers with sufficient funds, the probabilities and leaf-node value are combined in a deterministic equation thus:

$$\text{Probable income} = 0.9 \times 0.8 \times 0.4 \times \$40\text{MM} = \$11.52\text{MM} \qquad (10.1)$$

This calculation results in a value of $11.52MM, indicating that our enterprise likely will exceed the $10MM corporate-headquarters-imposed economic hurdle. Such calculations give us some concept concerning whether, in this case, the proposed plant will be economically successful. Decision trees of this type can be invaluable in graphically outlining a problem and in determining, on a conceptual level, the likelihood that a proposed project will succeed or fail.

Decision trees, however, seem to reach their practical limit with the conceptual model. That is, when copious quantities of data, complex probabilistic branching, time-series analysis, etc., are introduced, the decision-tree approach becomes, at least, unwieldy. I use the term analytical (many other terms are used by myriad people to describe the same thing) to indicate a decision tree into which such information has been injected.

Consider, for example, the plant-building situation several months after the construction of the conceptual decision tree. Over the past few months, teams of experts in the fields of demographics, marketing, and operations have individually met to put some numbers to the conceptual model. In the old business school style of assigning ranges to variables, management has requested that each team come up with a reasonable maximum value, a mean or medium value, and a low value. The low-values branch for customer base and percent-of-customer-base captured on the analytical decision tree is to represent the chance of attaining a minimum success value. High values for operating costs are to have similar meaning.

In addition, each team is to fabricate a leaf-node value for each of the branches. To generate such values, representatives from each team have to meet to discuss the scenario represented by each branch. Again, in the old business school fashion, the teams are required to generate but a single value for each leaf node. (A bit later in this section, I will discuss the use of distributions in conjunction with decision trees.)

Figure 10.4 shows the analytical decision tree for this project. I mentioned when describing the conceptual decision tree above that the reason for limiting this example to just three variables would become clear when an analytical tree was built for the project. It should be painfully obvious that this methodology is not an expandable

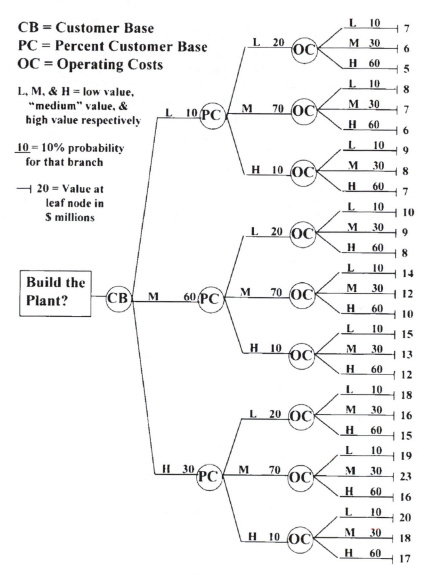

CB = Customer Base
PC = Percent Customer Base
OC = Operating Costs

L, M, & H = low value,
"medium" value, &
high value respectively

<u>10</u> = 10% probability
for that branch

—| 20 = Value at
leaf node in
$ millions

FIGURE 10.4 Three-variable analytical decision tree.

architecture. The tree with three variables already has 27 leaf nodes to which values have to be ascribed. Addition of just one more variable raises the number of leaf nodes to 81.

I have performed numerous risk assessments for projects like the one described here. Considering technical, scientific, engineering, commercial, environmental, financial, political, and other component parts of such a risk model, it is not uncommon for this type of model to be composed of 50 or more variables. I have found it to be impractical to use the decision-tree type of analysis on real-world risk models. A decision tree turns into a decision bush.

In addition to the proliferation of branches, I find it simply unreasonable to expect anyone or any team to be capable of (or have the time for) talking through the logic of each branch and ascribing a value to each leaf node. For the tree shown in Figure 10.4, teams would be required to talk through 27 logic trains. Addition of a fourth variable commits them to discuss 81 separate scenarios.

Aside from the node-proliferation problem, most real-world risk assessments of this type consist of components and processes that are not well represented by the analytical decision tree approach. Severely branching logic (IF and ELSE statements), looping (WHILE statements), time-series analysis (considering the third dimension of time), and other commonly employed risk-model building techniques I find to be all but excluded by the analytical decision tree approach.

Solving the tree is elementary. A decision-tree practitioner need only sum the products of each branch. The tree shown in Figure 10.4 would be solved by the following system of equations.

$$\text{Plant profit} = 0.1 \times 0.2 \times 0.1 \times \$7\text{MM}$$

$$+ 0.1 \times 0.2 \times 0.3 \times \$6\text{MM}$$

$$+ 0.1 \times 0.2 \times 0.6 \times \$5\text{MM}$$

$$+ 0.1 \times 0.7 \times 0.1 \times \$8\text{MM}$$

$$+ 0.1 \times 0.7 \times 0.3 \times \$7\text{MM}$$

$$+ 0.1 \times 0.7 \times 0.6 \times \$6\text{MM}$$

$$+ 0.1 \times 0.1 \times 0.1 \times \$9\text{MM}$$

$$+ 0.1 \times 0.1 \times 0.3 \times \$8\text{MM}$$

$$+ 0.1 \times 0.1 \times 0.6 \times \$7\text{MM}$$

$$+ 0.6 \times 0.2 \times 0.1 \times \$10\text{MM}$$

$$+ 0.6 \times 0.2 \times 0.3 \times \$9\text{MM}$$

$$+ 0.6 \times 0.2 \times 0.6 \times \$8\text{MM}$$

$$+ 0.6 \times 0.7 \times 0.1 \times \$14\text{MM}$$

$$+ 0.6 \times 0.7 \times 0.3 \times \$12\text{MM}$$

$$+ 0.6 \times 0.7 \times 0.6 \times \$10\text{MM}$$

$$+ 0.6 \times 0.1 \times 0.1 \times \$15\text{MM}$$

$$+ 0.6 \times 0.1 \times 0.3 \times \$13\text{MM}$$

$$+ 0.6 \times 0.1 \times 0.6 \times \$12MM$$

$$+ 0.3 \times 0.2 \times 0.1 \times \$18MM$$

$$+ 0.3 \times 0.2 \times 0.3 \times \$16MM$$

$$+ 0.3 \times 0.2 \times 0.6 \times \$15MM$$

$$+ 0.3 \times 0.7 \times 0.1 \times \$19MM$$

$$+ 0.3 \times 0.7 \times 0.3 \times \$23MM$$

$$+ 0.3 \times 0.7 \times 0.6 \times \$16MM$$

$$+ 0.3 \times 0.1 \times 0.1 \times \$20MM$$

$$+ 0.3 \times 0.1 \times 0.3 \times \$18MM$$

$$+ 0.3 \times 0.1 \times 0.6 \times \$17MM = \$12.049MM \qquad (10.2)$$

The tree's resultant value of $12.049MM exceeds the hurdle value of $10MM set by corporate headquarters. In fact, it exceeds the conceptual tree estimate of $11.52MM. So, the project likely will go ahead.

Some of the sting of employing analytical decision trees can be abated by the use of modern software packages that allow the user to represent each leaf node with a complete distribution. This technique may or may not reduce the number of branches and nodes (depending on how it is employed). It does, however, avoid the problem of discretizing (selecting a small and fixed number of values to represent an entire distribution) which, given modern software and statistics-free distribution-building technology, is never necessary or even desirable.

Another practical approach to addressing analytical decision trees that have grown into decision bushes is to simply convert them to a Monte Carlo-type model. This approach will sacrifice the graphical tree representation of the problem (the graphical tree representation can hardly be considered an advantage in a many-variable tree) but does afford many advantages. Using entire distributions instead of a limited number of discrete values is facilitated. This process allows the model to retain all of the relevant data. In addition, functions such as time-series analysis, branching, looping, and other processes are easily accommodated. See the example Decision-Tree Conversion in Chapter 17 for an illustration of such a conversion.

FACTOR ANALYSIS

Of the technologies presented here, factor analysis is the most difficult for people to conceptualize and understand. This, of course, means that it is difficult to explain. However, I am obligated to take a whack at it, so here goes.

The fundamental purpose of factor analysis is to reduce the dimensions of a problem. This means that factor analysis is good at reducing the dimensionality of the variable space in which data points are plotted. For example, we may have accessed the census bureau files on a group of people. The files contain a large number of variables (let's say 100) measured on each person. If you were to plot each person in 100-dimensional space and could view the plot, you might see some grouping of people (clustering, if you will) but would be hard pressed to tell anyone which variables contributed significantly to the clustering of bodies. Factor analysis helps you to reduce the dimension space (from 100 to some manageable number) by creating a new set of uncorrelated variables that contain most of the information contained in the original set of 100 variables.

Geometrically speaking, uncorrelated variables and their axes are those, in a 3-D world, that when plotted have axes at right angles to one another. One of the mystifying aspects of statistics is that you can have as many axes at right angles (uncorrelated) as you like. For us 3-D thinkers, it bends our gray matter all out of shape to try to imagine such a thing. Factor analysis is steeped in jargon. Some of the more esoteric terms are eigenvector and eigenvalue. These relate to the uncorrelated axes discussed in the previous paragraph. Let me explain how.

Imagine a cloud (like you see in the sky) floating in space. We can only imagine 3-D space (not counting time), but remember that in the statistical world, this cloud could be floating in n-space. All the particles which comprise the cloud can be thought of as data points plotted in n-space. Let us assume that n is some unmanageable number of variables, and we would like to boil down the number of variables to a smaller, more comprehensive, and representative set of parameters.

Imagine now that you are standing inside the cloud and have with you a measuring tape. This tape can get longer, but cannot bend, so you can only measure distances along straight lines (in 3-D space). What you would like to know is, how long is the longest straight line you can draw within the confines of the cloud? Through a process that I will not describe here, factor analysis facilitates the discovery of this longest line through the cloud. A line such as this is called an eigenvector and its length is called the eigenvalue for that eigenvector. This, loosely, can be considered the first principal component in factor-analysis talk.

This longest eigenvector represents the direction through the cloud that describes the greatest amount of variability in the shape of the cloud. It describes more variance in the plotted data than any other line you can draw. This line or eigenvector is represented by an equation, the terms of which are all the original variables.

The first eigenvector may account for, say, 55% of the variance of the dataset. Because we want a new set of variables that describe most of the variance represented by the original set (n) of variables, we should not be satisfied with 55% and will seek to find a second eigenvector.

The second eigenvector is the second-longest line you can draw through the cloud, with the catch that this second line must be at right angles (again, the term right angles only is good in a 2- or 3-D world) to the first eigenvector; i.e., it is uncorrelated. Again, factor analysis facilitates finding this axis, which describes the second greatest amount of variance. This axis may account for, say, 30% of the variance in the dataset. Now we have accounted for 85% of the variance. The identification of about two more eigenvectors will likely bring us to a percentage variance value greater than 95%. In the world of factor analysis, the contributions of additional eigenvectors becomes increasingly small, so it rarely pays to define more than the number of vectors that it takes to describe 95 to 99% of the variance.

When factor analysis is through finding axes through the cloud, we have a much reduced number of variables that do a good job of describing what this cloud looks like (describing the data distribution). Well, what in the world does this have to do with risk assessment?

Each original variable will be a term in the equation of each eigenvector. Let us say that we have a cloud floating in 3-D space (X, Y, Z) and that its longest dimension is nearly parallel to one of the original axes — let us say the X axis. It stands to reason that the longest line that one can draw through the cloud will be nearly parallel to X. This means that X will be a very important or influential variable in the equation that describes the nearly X-parallel eigenvector. The measure of how important a variable is to the equation that describes the eigenvector (how parallel the variable axis is to the eigenvector) is called a loading for the variable. Each original variable has a loading value for an eigenvector. These loadings can be used to determine which variables are the most important in a given dataset. This can be very significant information when you are attempting to perform a risk analysis but you are faced with a large set of variables, and you have no idea as to which ones you should pay most attention. The loading values can be thought of as weightings for the variables for use in risk-weighted calculations. This kind of analysis can be exceedingly helpful when attempting to determine for a risk analysis which are the variables that are most likely to bite you. The risk analysis, however, may actually be run on the newly defined (and smaller number of) factors.

Now that I have played up this technology, let me deflate it a bit by mentioning an associated practical problem. Eigenvectors are great things and are powerful data analysis tools, but when you use them as the variables in an analysis and your boss asks you what they mean (i.e., "Glenn, what is the significance of factor 2 having such a high value in the final analysis?"), you had better be prepared to describe just what is factor 2. This often is difficult because factor 2, like all other factors, is a combination of all the original variables. If you are lucky, factor 2 will have high loadings for only one or two of the original variables. However, it often is the case that many original variables have significant contributions to factor 2 (high loadings for factor 2), and it becomes practically impossible to describe to the uninitiated just what is the meaning of a change in factor 2. This explanation becomes even more difficult when the relative changes of multiple factors have to be explained. What this means in a practical sense is that factor analysis is a very powerful data analysis and risk assessment tool, but is not much of a presentation vehicle.

NEURAL NETS

As the name would imply, neural nets are computer systems that try to emulate the function of the brain. Neural nets (NNs) can be thought of as a branch of the more general sciences of artificial intelligence and expert systems. NNs, however, have special characteristics that make them not only unique, but fundamentally more powerful and generally applicable than many other artificial intelligence techniques or expert systems.

As already stated, the aim of NNs is to build a computer system that can process information like the brain. The fundamental elements of a brain are, to oversimplify the situation, the neurons, synapses, dendrites, and axions. Dendrites collect information from other brain cells and from the outside and communicate this chemical message through synapses to the neuron. The neuron is the basic cell in the nervous system that receives and combines information. Output from a neuron is carried to other cells, etc., by the axions. The human brain is composed of billions of these structures.

The fundamental elements of a typical NN are designed to emulate these basic nervous system elements. The processing element in an NN is the (rudimentary) equivalent of the brain's neuron. The processing element receives information from many input paths (dendrites and synapses) and internally combines the inputs and makes decisions about passing out output signals.

The information carried by an input path to a processing element usually is weighted. Multiple weighted pieces of information are summed in the processing element. The processing element also contains a transfer function. This function can be a simple threshold (i.e., if the sum of the weighted input information is not great enough, then no output is generated) or can be a complex, continuous function of the summed input. The output from the transfer function is passed to the output path (axion) which, in turn, can be used as input to subsequent processing elements. NNs are comprised of many such connections of processing elements. Typically, the processing elements are configured in layers.

One of the real advantages of NNs is their ability to learn. The learning process is fundamentally a matter of generating, for a given input, the desired output by adjusting the weighted input at the various processing elements. There are many types of learning (supervised, unsupervised, reinforcement, etc.), the details of which will not be conveyed here. Suffice it to say, NNs offer the ability to generate a desired response and to modify the response through time (and experience) without having to write a traditional computer program for each unique desired outcome.

NNs are excellent at such things as pattern recognition, signal processing, target classification, and many other tasks. One of the tasks to which NNs can be put is that of prediction. It has been shown that NNs, in certain cases, offer orders of magnitude improvement over more conventional linear and polynomial methods of predicting. Much of the use of risk assessment (Monte Carlo analysis, etc.) is involved with predicting the possible outcomes of a situation and the likelihoods associated with each probable outcome. In appropriate situations, NNs offer a powerful, practical, and flexible method of risk assessment.

Another area where NNs may become practical in the future is that of system modeling. Many of the risk assessments performed are of the process type. That is, they are not basically probabilistic in nature but rather attempt to emulate a natural process in order to evaluate the possible outcomes. Although probably not quite sophisticated enough yet to handle complex process models, NNs likely offer a flexible solution to relatively simple process situations.

THE RISK MATRIX

One of the most popular semi-quantitative methods of evaluating and displaying risk-related information is the use of risk matrices. A risk matrix is nothing more than a grid. The X and Y axes (usually, the bottom and left sides of the box) represent some measurable or estimable parameter whose severity, typically, increases from left to right on the X axis and from bottom to top on the Y axis.

A simple example of such a matrix is shown in Figure 10.5. In this matrix, I have created nine boxes, but keep in mind that a matrix can contain as many subdivisions as needed and need not have the same number of subdivisions in the X direction as in the Y direction.

The matrix shown in Figure 10.5 might be used for terrorism assessment for a physical facility. The Y axis is labeled Threat to Security. Things identified as high threats to security would plot near the top end of the Y-axis range while items identified as low threats would plot near the bottom of the Y axis. The X axis represents our estimated ability to counter any identified threat. If we are able to cope with the threat, then our ability to counter is high. If the threat represents something that we have little power to mitigate, then our ability to counter is low.

Threats that plot in the lower left corner of the matrix represent those threats that are less serious and that can be mitigated to a significant degree. Threats that plot near the upper right corner of this plot are those upon which we should focus our attention — these are significant threats about which we can currently do little to counter. Our efforts to secure our facility should mainly, but not exclusively, focus on these items.

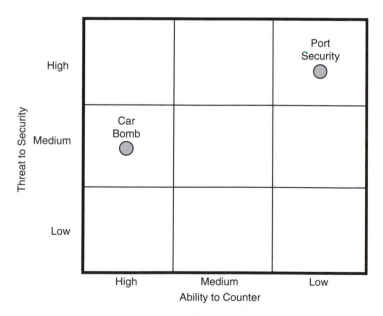

FIGURE 10.5 Risk matrix that might be used to evaluate threats to a facility.

For example, our facility might be positioned near a major sea port in a foreign country. Regulation of the port is exclusively the realm of the host government. In the past several years there have been numerous incidents of hazardous materials being mishandled by dock workers and it is generally known that chemical and biological weapons have passed through the port and into the hands of local terrorist groups. Our efforts to influence the government with regard to port security have, in the past, been met with ambiguity and sometimes disdain. We feel our facility is significantly threatened by the mishandling of hazardous materials and by the weapons being supplied to local terrorist organizations.

So, we would consider this threat to be high and our ability to counter — at this time — to be low. This threat then would plot near the upper right corner of the matrix. Our focus should be on renewed efforts to influence the host government to better regulate commerce at the port. Lobbying the U.S. government to bring pressure on the host government is one tactic that might be explored. In addition, we might raise the threat of moving our facility to another country if security is not improved. Our facility is a major source of jobs and tax revenue in the host country, and this economic leverage has yet to be exploited.

A second perceived threat is that of car bombs. Although there have been no incidents of such terrorist actions in the area, it is clear from inspection of the facility's physical layout that such an attack could have a relatively minor impact on personnel and production. The plant is ringed by a fence and low cement curbs, but these cursory barriers would not stop the determined individual. The main offices and plant facilities are well inside the boundary of the plant perimeter, so it is unlikely that any such car bomb could injure or kill workers or cause major damage to production facilities. However, a bomb set to go off at the time of shift changes could threaten many lives.

We might consider this threat to be of medium severity and we believe that we could, with relatively minor and inexpensive tactics and physical modifications, counter this threat. Therefore, this problem would plot near the center of the Y axis and near the left end of the X axis as shown in Figure 10.5.

As with all such matrix representations, it is not the plot that is of great benefit. The major positive impact of this risk assessment approach — as with nearly all other methods — is the conversations fostered regarding the risks and what can be done about them. Meaningful and in-depth conversations regarding the identification and mitigation of risk is always the salient positive ramification of implementing this, or most other, risk assessment processes.

One more word about the risk matrix approach. Some more popular incarnations of the matrix include the combination of multiple variables on one or both of the axes. For example, it is not uncommon to use the Y axis to represent the product of, say, impact and probability where impact might be measured in dollars and probability in percent. This approach has its benefits. One is that it forces people to quantify their assessments. This is typically a good thing because it causes more in-depth examination of the identified risks and fosters consensus among those involved in the process. This multiple-parameter approach also has a dark side.

Consider that we have identified a major threat — something that could cause significant harm to our project. Such a threat might have a very high impact (dollar) value. However, the threat is deemed not likely to occur. So, the probability of such a threat might be very low. Depending on the relative magnitudes of the impact and probability coefficients, this identified threat could end up plotting in a square in the matrix that would cause us to not focus on this problem. Some would argue that this is right because if it is a low-probability event, we should not focus on it. I would argue that because it is a high-impact event, it probably warrants more attention than a relatively low-impact and low-probability event that might plot in the same square. If we used a 3-D matrix (one axis probability, one axis impact, and one axis ability to counter), this event would clearly stand out and could be given appropriate attention. The point is, be careful when using a matrix in which any axis represents the combination of multiple parameters. In such a matrix, events that cluster together are not necessarily of the same stripe.

RISK REGISTERS

Risk registers are (or should be) relatively simple mechanisms for capturing risk-related data and, best of all, fostering conversations about the risks and what will be done about them. A simple example of a risk register is shown in Figure 10.6.

In this risk register, the first column is typically a description of the risks identified. In this case, the project team has thus far identified two risks. They are as follows:

- Contractor availability. The company might not in a timely manner be able to secure the contractor skills needed. Competition for contractor help is keen.
- Permits. In this part of the world, issuance of permits is often held up for political reasons.

The second column in the risk register might be a simple hyperlink. Clicking on this item would take the reader to a document, for example, that contained an in-depth discussion of the identified risk. Such a document is not shown here.

Actions designed to mitigate the identified risk could constitute the third column. This also might be a hyperlink to a more extensive discussion of the action to mitigate. In our simple risk register shown in Figure 10.6, early negotiation with potential contractors is identified as a possible action to mitigate the contractor availability problem. Early intervention in the permit issuance process is identified as an action that might be taken to offset the threat of delayed permits.

RISK	DEFINITION	MITIGATE ACTION	PERSON ACCOUNTABLE	DATE	MIN. $	M.L. $	MAX. $	PRIORITY
Cont. Avail.	Click here for full expl.	Early negotiation	Joe B.	02/05	$0.7K	$1.2K	$1.5K	H
Permits	Click here for full expl.	Process — early start	Sally S.	05/05	$1.2M	$1.5M	$1.7M	M

FIGURE 10.6 Example layout for a risk register.

The next two columns simply identify who is responsible for pursuing the actions to mitigate and by when they will have executed the tasks agreed upon. These are important parameters. Too often, risk-type meetings are held, risks identified, actions identified, and the meeting adjourns without accountability being assigned.

Costs are also a significant part of the risk-register conversation. The next three columns capture the range of potential costs to mitigate the problem. Consensus on this range is an important element of the process. For example, if the cost to mitigate is likely to be greater than the cost of the risk if it happens, project team members need to consider whether the cost/benefit ratio of this risk and action to mitigate is sufficient to pursue the mitigation step.

Finally, in this simple risk register, is an estimate of the priority of this perceived risk. Such assigned priority is not only a function of the risk itself, but is also a function of the timing of such a risk relative to the point in time of the risk-register conversation. For example, we might deem that we need to sign contracts with potential contractors right away. Any delay could cause the inception of the project to be delayed. Getting permits is also essential, but the project currently is in the planning stage and permits, though important, will not be required until well after other initial project steps and tasks have been executed. Therefore, in our risk register the team might assign — at this point in time — a high priority to contractor availability and a medium priority to the permit problem.

It is important to note that creation and population of a risk register is not a one-time event. As indicated in the preceding paragraph, priorities, costs, and other parameters change with time. A few weeks after this initial risk register was created, contracts with critical third parties might have been negotiated, thus dropping the priority of this risk to low. Because time has passed and we are into the initial stages of design of the project, the priority of permits might also be reevaluated.

Again, like the risk matrix approach or nearly any other risk-assessment process, the main benefit of such a method is the conversations that are fostered by implementing the process. The plots and charts are nice and do afford a method of recording items and tracking progress. However, it is the conversational interaction of all interested parties that is the primary benefit of any such risk-assessment action.

SELECTED READINGS

Bayes, T., An essay towards solving problems in the doctrine of chances, *Philosophical Transactions of Royal Society* 53, 1763.

Carbonell, J. G., Ed., *Machine Learning – Paradigms and Methods*, The MIT Press, Cambridge, MA, 1990.

Davis, J. C., *Statistics and Data Analysis in Geology, Volume 2*, John Wiley & Sons, New York, NY, 1986.

Harbaugh, J. W., Davis, J. C., and Wendebourg, J., *Computing Risk for Oil Prospects: Principles and Programs*, Pergamon Press, Tarrytown, NY, 1995.

McNamee, D., *Business Risk Assessment,* Institute of Internal Auditors, Altamonte Springs, FL, 1998.

Newendorp, P. D., *Decision Analysis for Petroleum Exploration*, Pennwell, Tulsa, OK, 1975.

Nilsson, N. J., *Principles of Artificial Intelligence*, Tioga, Palo Alto, CA, 1980.

Schmucker, K. J., *Fuzzy Sets, Natural Language Computations, and Risk Analysis*, Computer Science Press, Rockville, MD, 1994.

11 Monte Carlo Analysis

A BIT OF HISTORY

Monte Carlo analysis uses the process of simulation to achieve a range of solutions to a problem. This technique generally is used to solve problems for which the definition of specific solution equations to calculate a specific answer is either too complex or too cumbersome to be practical. In Monte Carlo analysis, typically (but not necessarily) input variables are represented as distributions of values (see Chapter 12 of this book). Values comprising these distributions represent the range and frequency of occurrence of the possible values for the variables. Also needed are equations that relate the input variables to the output (result or answer) variables.

Monte Carlo analysis is a statistical procedure based upon random selection or chance. The name Monte Carlo, of course, is taken from the city of Monte Carlo, Monaco, made famous by games of chance.

Monte Carlo analysis, or distribution sampling, was arguably first presented formally by Student (W. S. Gosset, inventor of the Student's t-test) in the early 1900s. At first, Monte Carlo analysis was treated as being distinct from experimental sampling or model sampling, both of which involve random selection from populations. The term Monte Carlo was restricted to processes in which weighted or importance sampling was used to increase the efficiency of the sampling process. In other words, the term Monte Carlo was reserved for sampling processes that embodied some smarts to increase sampling efficiency. This restricted use of the term eventually faded because high-speed computers lessened the importance of an efficient sampling process.

Today, the term Monte Carlo analysis can be applied to any procedure that uses distribution-based random sampling to approximate solutions to probabilistic or deterministic problems. These types of problems were exemplified by the first practical use of the Monte Carlo method — addressing the probabilistic aspects of particle diffusion associated with the design and construction of the first atomic weapons at Los Alamos during World War II.

WHAT IS IT GOOD FOR?

Most processes — physical, chemical, or otherwise — terminate in a result. Many processes are so complex that the generation of an equation or system of equations to calculate the result of the process is arduous or impossible. Monte Carlo analysis often is used to simulate these processes and to generate probable results. This avoids having to resort to the construction of specific equations which can predict the outcome of the process. In addition, Monte Carlo analysis can yield likelihoods for a range of possible results.

Although Monte Carlo analysis has many uses in business and science, the most common application is that of determining the probability that a certain event (or

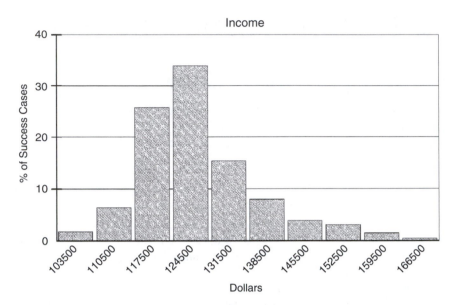

FIGURE 11.1 Distribution of income.

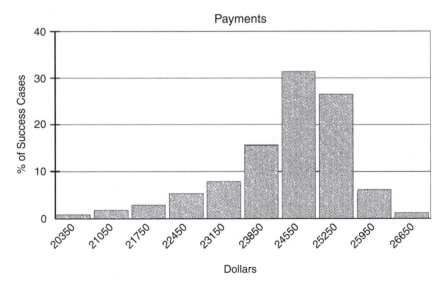

FIGURE 11.2 Distribution of payments.

result) will occur and getting a grip on the magnitude of the event. The fundamental technique used to accomplish these things is the random sampling of values from frequency distributions which represent the variables in the analysis (see Figure 11.1 and Figure 11.2). Well, what does all that mean?

Every risk assessment study involves the integration of at least several and sometimes many variables. Variables are the basic elements of the problem (porosity of the soil, concentration of an element in groundwater, distance from a potable water source, etc.). If Monte Carlo analysis is to be used, each of these variables usually is represented by a distribution of values (coefficients of the variable). The values that constitute the distribution should span the range of possible values for the variable in question.

Monte Carlo analysis, in its native form, also requires that the variables be independent. That is, they must not have a relatively high correlation. For example, if two variables in a Monte Carlo analysis are depth and temperature, then it is not valid to randomly select a depth and temperature value from their respective distributions and expect that this combination will emulate a real situation. This is because, in general, as depth increases, temperature increases. A randomly selected depth of 5 feet and a temperature of 500°F do not, in most cases, simulate real-world conditions. To account for the correlation between variables, the technique of defining dependence between the two variables should be employed. See Chapter 14 of this book for an explanation of this technology.

SIMPLE MONTE CARLO EXAMPLE

For the sake of an example of Monte Carlo analysis, let us consider the oversimplified example of calculating profit from income and payments. (Payments are salaries, taxes, and other costs.) Figure 11.1 shows a distribution of possible incomes for the year. Figure 11.2 shows the range of payments we make in the same year. On each iteration of Monte Carlo analysis, a randomly selected income will be combined with a randomly selected payment to yield a profit value using, for example, the following equation:

$$\text{profit} = \text{income} - \text{payment} \qquad (11.1)$$

The plot of profits shown in Figure 11.3 is the result of 1500 random selections and profit calculations. This model, of course, assumes that there is not a relationship between income and payments. This is not likely. See Chapter 14 for a solution to this dilemma.

BUCKETS

In the chapter on distributions, I will cover in depth what distributions are, how to think about them, the various types of distributions, and more. This is one of those chicken-and-egg situations in which I need something now that I do not cover in detail until later in the book. Because it makes more sense overall to delineate distributions in the chapter following the Monte Carlo chapter, I will here give a cartoon version of the distribution discussion. Please see the next chapter for an in-depth discussion of the subject.

I find when teaching classes that a picture really is worth a thousand words. When discussing how Monte Carlo analysis actually works, I typically make an initial drawing like that shown in Figure 11.4. In this depiction of a distribution,

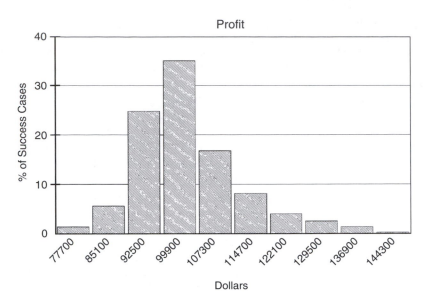

FIGURE 11.3 Distribution of profit resulting from Monte Carlo combination of income and payments.

each labeled bead represents a potential value for cost. One distribution displays our estimates for company labor costs and the other depicts our assessment of contractor labor costs.

In a computer-generated distribution, the X axis typically is comprised of 100 or more unique values. Only five values are shown for the sake of drawing convenience. The bar widths are arbitrary. Fewer or more bars can be created by respectively increasing or decreasing the bar widths. Bar heights represent, typically, the number of X-axis values captured in the bar width (as in this case), or, the percent of such X-axis values. The bars, however, are superfluous with respect to the distribution. The distribution itself is actually only the scattering of values on the X axis — as shown in Figure 11.5. The beads can be thought of as having been scraped off the X axis of the distribution and dumped in a bucket. The beads are then jumbled up and one bead is randomly drawn from each bucket.

In this case, we might be attempting to calculate the total labor cost which will be the sum of company labor cost and contract labor cost. If we were to perform a Monte Carlo analysis of 1000 iterations, we would on the first iteration draw a bead at random from each bucket, add the values we drew, record the sum on the X axis of our total labor cost plot, and return the beads to their respective buckets.

On the second iteration, we would repeat the process. This would be repeated 1000 times resulting in 1000 sums plotted on our total labor cost X axis. This, then, is the distribution of total labor costs. This example is, admittedly, a bit cartoonish, but it does seem to convey in a simple manner the concept of Monte Carlo analysis and how it interfaces with distributions.

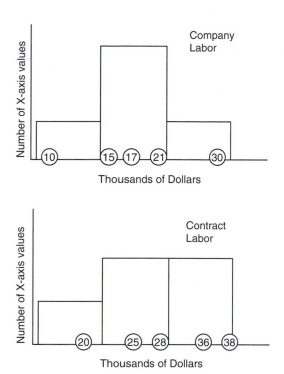

FIGURE 11.4 Cartoon depiction of a distribution. Each numbered bead is a unique value in the distribution. One array of values represents company labor costs and the other contractor labor costs.

HOW MANY RANDOM COMPARISONS
ARE ENOUGH?

The object in Monte Carlo analysis is to perform a sufficient number of comparisons from the input variable distributions so that the resulting variable distribution is valid. Some risk assessment packages allow the user to specify a convergence value that automatically will stop iterations when comparisons of mean values (or other selected values) from multiple simulations are less than or equal to the assigned convergence value. However, the number of distributions, the complexity of the equations, and the familiarity of the user with the situation may preclude his knowing such a convergence value prior to the analysis.

A simple method for determining whether a sufficient number of comparisons have been made is to inspect the mean value of the answer variable distribution from multiple Monte Carlo analyses. For example, if we have two input variables called X and Y and our equation is

$$Z = X + Y \tag{11.2}$$

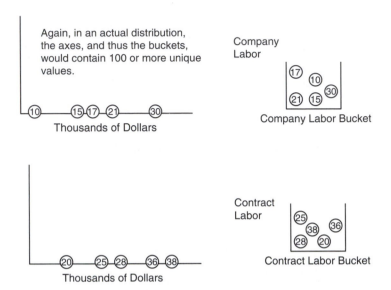

FIGURE 11.5 Distribution of labeled beads. For use in Monte Carlo analysis, it is convenient to think about the beads being put into a bucket and jumbled up. Random draws are made from buckets.

then the Z distributions from multiple individual Monte Carlo analyses (an individual Monte Carlo analysis may be comprised of, say, 5000 iterations) would be inspected. If the means of the individual resulting distributions do not vary significantly (significance is determined by the user), then it is likely that a sufficient number of comparisons has been made. Statistical tests for the significance of the difference of population means (t-tests) or of the entire distribution using the Kolmogorov–Smirnov test are available but can be complicated. Most of the time, simple inspection of the change in the mean of the resultant distributions and the shape of the distributions will suffice.

OUTPUT FROM MONTE CARLO ANALYSIS — THE FREQUENCY AND CUMULATIVE FREQUENCY PLOTS

As depicted in Figure 11.3, the primary output vehicle for Monte Carlo analysis is the frequency plot. The number of bars in a frequency plot can be set by the person generating the model. Bar heights simply indicate the number (or percent) of X-axis values that fall within the X-axis range dictated by the width of the bars.

Although useful, frequency plots are limited in the amount and type of information conveyed. In risk assessment and portfolio management, it is more common to view Monte Carlo output in cumulative frequency space. The same data shown in frequency space in Figure 11.3 is depicted in cumulative frequency space in Figure 11.6.

The conversion of frequency space to cumulative frequency space can be conceptually viewed as stacking the bars of a frequency plot. Figure 11.7 shows a four-

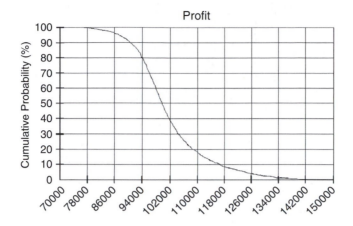

FIGURE 11.6 Cumulative frequency plot equivalent to frequency plot in Figure 11.3.

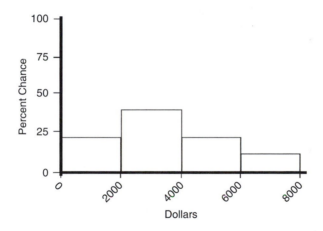

FIGURE 11.7 Four-bar frequency plot.

bar frequency plot. In Figure 11.8, the same four bars are stacked from right to left. In Figure 11.9, a dot has been inserted at the top of each bar and the dots connected with a line. In Figure 11.10, the bars have been removed, leaving just the cumulative frequency curve.

In the computer, of course, there actually are no bars. Software systems that build cumulative frequency curves work directly with the individual X-axis values generated by Monte Carlo analysis. Figure 11.11 shows a plot of eight values resulting from an eight-iteration Monte Carlo risk assessment model. The eight values have been sorted from smallest (values of least magnitude) on the left to largest (values of greatest magnitude) on the right.

Because eight values were generated, the cumulative frequency algorithm will divide the Y axis into eight equal probability segments. The largest value is moved up one probability segment. The second largest value is moved up two probability segments and so on until the algorithm reaches the last data point (leftmost or

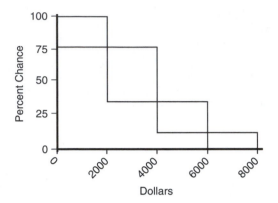

FIGURE 11.8 Four bars from Figure 11.7 stacked.

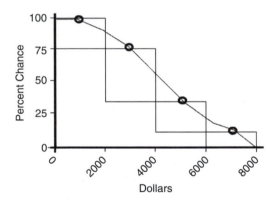

FIGURE 11.9 Curve connecting tops of bars from Figure 11.8.

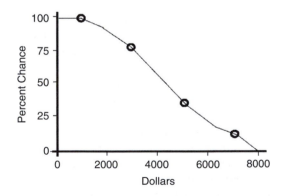

FIGURE 11.10   Cumulative frequency plot equivalent to frequency plot in Figure 11.7.

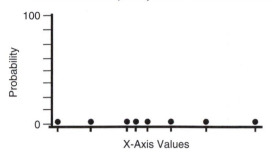

FIGURE 11.11 Results from an eight-iteration Monte Carlo analysis.

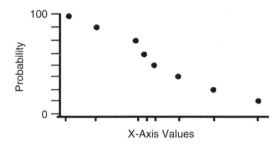

FIGURE 11.12 Individual X-axis values raised parallel to the Y axis according to their cumulative probability.

smallest X-axis value), which is moved up, in this case, eight probability segments. The points are connected with a smooth line which is the cumulative frequency curve (Figure 11.12 and Figure 11.13).

INTERPRETING CUMULATIVE FREQUENCY PLOTS

Well, now that we have constructed a cumulative frequency curve, what good is it and how do we read one? Frequency can be accumulated to the left or to the right, and the resulting Z or S curves, respectively, impart similar information. However, there is a slight difference in the interpretation of the two curves.

In Figure 11.14 and Figure 11.15, both curve types are shown. The X axis for both curves is profit in dollars; the Y axis is probability. The dashed lines on each plot connect a dollar amount with a probability. The dashed line in Figure 11.14 is interpreted to mean that there is a 45% chance of attaining a profit of about $570 or greater. Or, stated differently, there is a 45% chance of getting at least $570 of profit. In Figure 11.15, the dashed line would be interpreted to indicate that there is a 65% chance of getting up to a profit of about $400. Stated another way, there is a 65% chance of obtaining a profit of $400 *or less*.

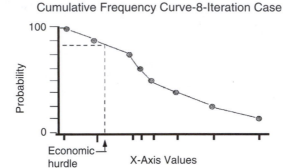

FIGURE 11.13 A cumulative frequency curve constructed by connecting the points in Figure 11.12.

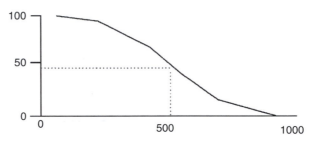

In this case, we can say that we have about a 45% chance of getting "at least" $550 or more.

FIGURE 11.14 Example of a Z-shaped cumulative frequency curves.

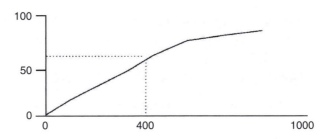

In this case, we can say that we have about a 65% chance of getting "up to" 400 (including 0).

FIGURE 11.15 Example of an S-shaped cumulative frequency curve.

People who deal with situations where bigger is better prefer the accumulating-to-the-left (Z-shaped) plot. For example, a petroleum engineer charged with estimating oil production from an oil well would want to know that the well will produce X amount of oil or more (or, at least X amount of oil). A cost accountant, on the other hand, might prefer the accumulated-to-the-right curve (S-shaped) because in

Variable A	Variable B
3	7
5	2
6	13
2	4
12	5
4	10
8	9
13	11
10	6
9	12
7	8
11	10
4	3
9	10
12	9
5	11
8	4
10	7
9	3
11	5

FIGURE 11.16 Two 20-value arrays for variables A and B.

her job it is important to know that a project has a Y percent chance of costing up to X dollars.

Most off-the-shelf risk assessment packages offer the S curve as the default situation. However, those with experience in all types of risk assessment prefer the Z curve because its high-probability end is the end near the Y axis. This has great advantage when incorporating chance of failure into a risk assessment (see Chapter 13), when calculating risk-weighted values (see Chapter 15), or when performing a host of other risk assessment tasks.

A typical misconception associated with cumulative frequency plots is that the low-value end of the curve (the end nearest the Y axis) should approach or reach zero. Consider the situation of two distributions, A and B, displayed as arrays of values in Figure 11.16. In each distribution, the smallest value is 2, and the value 2 only occurs one time in each distribution. The risk model that uses distributions A and B is simply using Monte Carlo random sampling to add A and B to produce C (C = A + B).

The value of least magnitude that C can attain is 4 (A = 2 and B = 2). Given that there are 20 values each in the A and B distributions, we have a 1 in 400 (20 × 20) chance of producing a value of 4. You will note, however, that the value of 4 plots at 100% on the Y axis in Figure 11.17. It should be apparent that there is not a 100% chance of producing a value of 4 by random selection from A and B (only 1 chance in 400). So, what do we have a 100% chance of producing with the equation C = A + B? There is a 100% chance of calculating a C-value of 4 or greater. Other

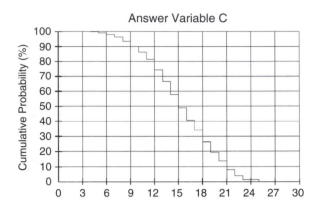

FIGURE 11.17 Cumulative frequency plot with a minimum value of 4 and a maximum value of 26. Curve created by summing variables A and B (Figure 11.16).

X- and Y-axis pairs are similarly interpreted (a Y probability of getting X or greater than X).

This leads to another point concerning the interpretation of cumulative frequency curves. In Figure 11.17, the value of least magnitude on the plot is also the value of least magnitude that could have been produced from the distributions A and B. Similarly, the maximum value on the curve is 26 that also is the value of maximum magnitude that could be calculated from distributions A and B (given our C = A + B equation). It is not typical that the end points of a cumulative frequency curve represent the minimum and maximum values that could have been produced.

Distributions A and B in Figure 11.16 each are comprised of only 20 values. If 400 or more Monte Carlo iterations are used to randomly sample distributions A and B, then there would be a pretty good chance of producing the minimum and maximum values that could result from adding A and B (the greater the number of iterations, the better the chance of producing the minimum and maximum values). Distributions in actual risk-assessment models typically contain 100 to 500 or more values. Also typical is to have many more than two distributions in the risk assessment model. Consider a model composed of, say, two distributions, each containing 200 values. We would need to run tens of thousands of Monte Carlo iterations to have any reasonable chance of producing a cumulative frequency curve that had as its end points the values of largest and smallest magnitude that could have been produced by combination of the two distributions.

Because of the common complexity of risk assessment models, the speed of the typical computer, and other factors, most risk assessment models are allowed to run a number of Monte Carlo iterations that is between 1,000 and 10,000. (The number of iterations generally can be set by the user and can be highly variable.) This is far too few iterations to ensure that values of minimum or maximum magnitude have been generated. Therefore, the end points of most cumulative frequency curves do not represent the values of minimum and maximum magnitude that could have been produced. The end points represent the values of minimum and maximum magnitude that were produced in the given number of iterations.

At first blush, this would appear a foible in the Monte Carlo process, but it is not. To illustrate the reason why you likely do not want absolute minimum and maximum values represented on the plot, consider the following situation. Suppose we are calculating profits from distributions representing number of units sold, price per unit, number of sales personnel, salary per hour for sales personnel, capital expenses, and operating expenses. If each of these distributions truly represents the range of values that could possibly occur (i.e., tails of the distributions are composed of values that are extreme and unlikely), there is a very small chance that, in the real world, there will be a simultaneous occurrence of the maximum number of units sold, the maximum price per unit, the minimum number of sales personnel, the minimum salary per hour for personnel, the minimum capital expense, and the minimum operating expense to produce the maximum possible profit. You may argue that it would be useful to calculate the minimum and maximum possible profit. I would counter this argument with the suggestion that you make such calculations on the back of a brown paper bag and that you do not need a risk assessment or Monte Carlo analysis to generate such values.

The purpose of Monte Carlo simulation, among other things, is to produce plots of things that are likely to happen. Alignment of absolute minimums and maximums, in the real world, is not likely. It also can be dangerous to produce such values.

Consider the situation of a company composed of several sales locations. The company would like to calculate the total combined profit from all stores. If each of the store-profit curves were allowed to range from the absolute minimum to the absolute maximum value (biggest and smallest values that could be produced), summation of these individual curves, even by Monte Carlo analysis (the way they should be summed) will generate upside and downside values that are highly unlikely and extreme. These values may imbue management with unrealistic expectations. (These numbers tend to get recorded in accounting sheets and reported.) These numbers also can be downright scary.

COMBINING MONTE CARLO OUTPUT CURVES

In the example in the immediately preceding paragraphs, we considered the problem of combining (in this case, combining by summing) the Monte Carlo output profit curves from individual stores to arrive at a total-profit curve for all stores. It is tempting (and I see it done frequently) to add together all of the respective individual output curve minimums, means (or medians), and maximums. Another typical mistake is to sum (or otherwise combine) percentile values such as P10, P50, and P90 values.

The reason that it is invalid to combine curves in this fashion is the same as that previously put forward regarding the combination of input distributions to Monte Carlo analysis and the calculation of values of absolute minimum and maximum magnitude. It is highly unlikely that extreme values will, in the real world, align to produce a resultant (sum) value of minimum or maximum magnitude. To correctly combine Monte Carlo output curves, each of the output curves should be used as an input curve to a new Monte Carlo routine specifically designed to combine the curves. When this approach is implemented, real-world and reasonable combinations and ranges result.

SELECTED READINGS

Bernstein, P. L., *Against the Gods — The Remarkable Story of Risk*, John Wiley & Sons, New York, NY, 1996.

Dore, A. G. and Sinding-Larsen, R., Eds., *Quantification and Prediction of Hydrocarbon Resources*, The Norwegian Petroleum Society (NPF), Special Publication No. 6, Elsevier, Amsterdam, 1996.

Gentle, J. E., *Random Number Generation and Monte Carlo Methods, 2nd Edition*, Springer-Verlag, New York, NY, 2003.

Jaeckel, P., *Monte Carlo Methods in Finance*, John Wiley & Sons, Hoboken, NJ, 2002.

Meyer, H. A., Ed., *Symposium on Monte Carlo Methods*, John Wiley & Sons, New York, NY, 1954.

12 Decisions and Distributions

DECISIONS

The only time we have to make decisions is when we do not know something about the situation. Consider the scenario in which your spouse asks you to go to the store. You leave the house and arrive at the proverbial fork in the road. You look down the right fork and see no store. You look down the left fork and you see the store. In this case, I submit that there is no decision to be made — the store is on the left fork and that is where you will go. However, if you reach the fork in the road and do not see a store down either fork, then a decision has to be made. Even if you decide to stay where you are or go back home, a decision still has to be made as to what to do. A decision has to be made because you do not know something. That is, you are uncertain about something related to the problem.

Distributions are one means of expressing uncertainty in a problem, and, in fact, using that uncertainty to our advantage. Therefore, distributions are used in decision making by affording us a means of quantifying, expressing, and using our uncertainty in the process of making a decision. At the end of this chapter, I will revisit the subject of decision making with a focus on flexibility in the decision-making process.

JUST WHAT IS A DISTRIBUTION?

One of the most common mechanisms for expressing uncertainty is the distribution. Just what does the term distribution mean? When you look at a distribution, just what should it be telling you? Consider the situation of the new-hire engineer who has been summoned on his first day of employment into the office of the head of construction. The engineer is informed that as part of the plant-expansion construction, 2000 doors and doorjambs will need to be built. The boss was informed by the supplier of materials that the company will have to buy its lumber in bulk directly from the mill. This means that the lumber will not be custom cut, but rather is available from the mill only in 1-foot increments. That is, the boards will come in 5-foot lengths or in 6-foot lengths or in 7-foot lengths, etc. The boss was unsure concerning what length lumber should be purchased, so she asked the engineer to measure the heights of several hundred adults in the community and present her with the measurement results so a lumber-length decision can be made.

Well, it is the engineer's first day on the job and he is anxious to impress the boss. In his zeal, he visits the local university and arranges to borrow some laser equipment used to measure distances. The engineer sets up the equipment to measure the heights of adults. The equipment is capable of measuring distances (heights) with great precision — out to 10 decimal places if needed. After a week of collecting data, the engineer thinks he has amassed a sufficient number of measurements and produces a plot of the heights. A plot of a subset of the data appears in Figure 12.1.

The measurements are carried out with sufficient precision that there are no repeating values; that is, each value is unique. So, the frequency with which each value occurs is 1. The plot of the data that the engineer takes to the boss is shown in Figure 12.2.

The boss takes one look at the plot and informs the engineer that the plot with which the engineer has presented her is not quite what she had in mind. She informs the engineer that she was expecting a more typical frequency plot with the bars of various height, and she sends him away to have another try.

"Well," the engineer wonders, "what is wrong with the plot I made?" After all, in a data set of nonrepeating values, each value occurs once (the probability of any value occurring is one divided by the number of values in the data set). While this seems perfectly reasonable to the engineer, he realizes that he will have to produce a plot that the boss will accept. To do this, he decides to change his way of thinking about the precision of the data he was presenting.

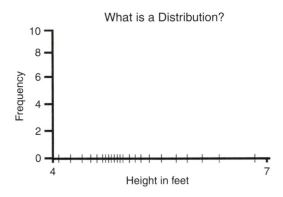

FIGURE 12.1 Measured values sorted and plotted on the X-axis.

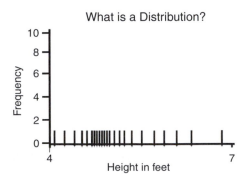

FIGURE 12.2 Plot of X-axis values each represented by a bar of height 1.

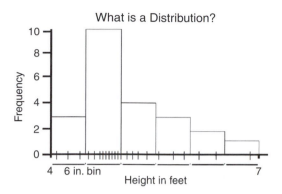

FIGURE 12.3 Plot of binned height data.

Rather than considering each value individually, he tumbles to the conclusion that he will have to consider the measurements in bins or buckets or groups. Figure 12.3 is the plot which results from considering the data in 6-inch-sized bins.

The boss is happy with the plot of heights so the engineer is happy too, but something still bothers him about the number of bins thing. He decides, however, to forget about it for the time being because the boss has given him a new assignment.

The new task is to analyze some financial data. He is to get from the accountants data that show the range of profits the company, from preliminary estimates, might make from the plant expansion. The question to be answered is, How likely is it that the company will realize, in the first year of production, a profit of around $5MM (the first-year economic hurdle for the project)? After retrieving the data from the accountants, the engineer loads the data into statistical software and makes three plots. These three plots are shown in Figure 12.4.

When five bins are used to represent the data, the bar chart has a roughly log normal appearance; that is, the plot is skewed to the right (remember, skewness direction is the direction in which the tail goes). When 10 bins are used, the resulting plot indicates that there is a wide range of values of about the same frequency with less frequently occurring values on either side. When 40 bars are used, the plot takes on a bimodal appearance.

The changing appearance of the plot, however, is not the most disturbing aspect of the plots. The boss wants to know what is the probability of generating a profit of about $5MM. According to the five-bar plot, the probability of making about $5MM is about 50%. According to the 10-bar plot, the probability is about 30%, and the 40-bar plot indicates that the probability is no greater than 9%.

It is clear from the plots that this assignment must be more clearly defined. Just what did the boss mean by about $5MM? If she meant the chance of achieving exactly $5MM, and $5MM just happened to be one of the X-axis values in the data set, then the probability of getting $5MM is 1 divided by the number of values in the data set. If $5MM is not one of the X-axis values plotted, then there is no chance of getting exactly $5MM.

However, the boss said about $5MM, so the engineer did not think she meant exactly $5MM. Without clarification concerning the definition of about, he could

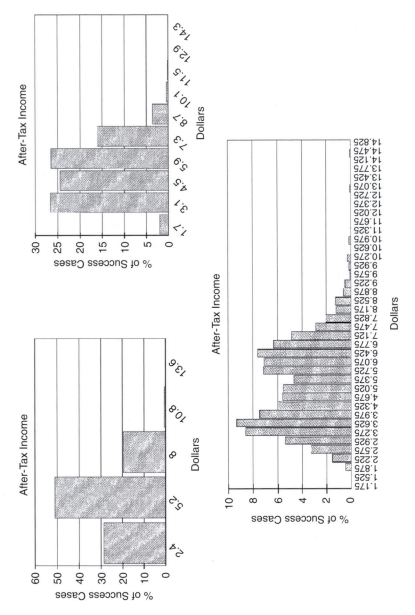

FIGURE 12.4 Frequency plots of common data using 5, 10, and 40 bins.

not determine the level of precision (i.e., bar or bin width) with which he should represent the data. It should be clear from the three bar charts in Figure 12.4 that the probability of achieving about $5MM depends on the bin size (bar width) which, in turn, is a function of the definition of about.

The lesson to be taken from this exercise is that when constructing a contributing factor diagram, the group building the diagram needs to discuss and arrive at a consensus decision regarding the level of precision with which the data need to be considered.

So, to return to the original question posed at the beginning of this section, just what is a distribution? The distribution actually is the scattering of data points along the X axis, as shown in Figure 12.1. A bar chart is a means of presenting the data, but it should be apparent that a presentation of the data should be commensurate with an understanding of the level of precision required.

For example, in our door-building example, because the lumber is cut in 1-foot increments (5 feet long, 6 feet long, etc.), it does not matter to the boss whether the average height measured is 5 feet 5 inches, or 5 feet 7 inches, etc. She makes the simple deduction that she needs lumber at least 6-feet in length. However, if the engineer's charge had been to make space suits for NASA instead of doors, this amount of imprecision in the data could not be tolerated (each suit is built specifically for each astronaut), and tolerances in height are very tight.

A bar chart, then, is a representation of the distribution. The X axis of the bar chart represents the range of possible values for the variable being graphed. X-axis values denote the probability associated with a given bin size. The bin size should represent the level of precision required. When the contributing factor diagram is being constructed, the group should discuss and come to consensus concerning the level of precision required and should agree that all plots made for the answer variable with respect this consensus. If this decision is not adhered to, very different representations of the data, such as those shown in Figure 12.4, can result. See the Output from Monte Carlo Analysis — The Frequency and Cumulative Frequency Plots section of Chapter 11 for information on converting frequency plots to cumulative frequency space.

DISTRIBUTIONS — HOW TO APPROACH THEM

Those versed in risk assessment or statistics tend to classify distributions by their properties. Such names as normal (or Gaussian), Poisson, gamma, and others are typical. Although this type of jargon is required if you want to be one of the boys and proliferate the erudite haze that shrouds statistics, I have seldom found such terms to be useful. Their limited use stems not so much from the fact that these names are relatively meaningless with respect to the actual distribution shape, but rather from the fact that my business has always been to bring risk assessment to people who are not necessarily versed in statistics or risk. Lawyers, accountants, plant managers, engineers, geoscientists, chemists, and others all are very well read with respect to the nuances of their given fields. Many of the people in these disciplines, however, can and do need to perform risk assessments. So while it is true that most of these people are capable in their fields, they likely are not intimately familiar with the fundamentals or particulars of statistics or risk assessment. I always liken it to driving a car. I think I can drive pretty well, but I really do not know much (anymore) about what goes on under the hood of my car. I count on someone else to take the complicated mess and make it easy for me to use, because I have to use the car.

That is my job with respect to statistics and risk assessment. Most people need to use these tools but, like the car, they really do not want to know what is going on under the hood, nor do they want to be asked to demonstrate a skill level or knowledge that is outside their ken. I like to say in my risk assessment classes that it is easy to make something hard, but it is really hard to make something easy. So it is with statistics and risk assessment.

I learned early in my career that if I were to be successful in bringing risk assessment to the masses, then I would have to find a way to insulate and isolate others from the rigors and details of statistics. This required not only an intimate knowledge of the subject matter on my part, but the development of algorithms, processes, and techniques that would allow those not steeped in statistics to use a risk assessment package and get the job done without having to provide unnatural (for them) information or to demonstrate extraordinary skills. I also learned early on that it is far better to use your intuition, experience, and knowledge of the subject to build a distribution shape that you think best represents the data or variable. Representing a variable's data by algorithm (i.e., selecting a gamma, or beta, or chi-squared distribution, etc.) is likely to be less useful. In most instances, users should be allowed to enter the data that they know about and should be able to shape the distribution to fit their mental model of the actual (parent, if you like) distribution of data. For example, an accountant may know (or think he knows) the distribution of product prices for the next 6 months. Rather than being asked to provide an alpha1 and alpha2 value or some combination of mean and standard deviation, variance, kurtosis, or any number of other (to them) meaningless statistical values, the user should be able to input data with which he is familiar (price data, in this case). He should be able to shape the distribution to fit historical price data, price projections, or experience-driven intuitive knowledge concerning the distribution shape. The algorithms I developed to build distributions do just that without bothering users with statistical terms.

The first step toward statistics-free risk assessment was the decision that I should strive to develop algorithms that would allow the generation and presentation of distributions by their general shape instead of by their statistical names. The names usually are tied to their properties (or inventors). To this end, algorithms were designed to allow the user to enter some combination of minimum (min), most likely (ml), maximum (max), and peakedness values that would combine to create (mostly) tailed distributions. The names of the distributions denote the distribution shape. The primary distributions for which algorithms were developed are as follows:

- Symmetrical
- Skewed
- Spike

- Flat
- Truncated
- Discrete
- Bimodal

Reading data (and randomly sampling the data) from a file, peakedness, and, as I like to refer to it in my classes, the evil, devil's triangular distribution will also be discussed in this section. I have taken to referring to the symmetrical, skewed, spike, flat, and truncated distribution types as simple distributions, not because their mathematics or forms are simple, but because simple methods have been devised for constructing these distributions. The distribution-building algorithms generated can build any of the aforementioned simple distribution shapes using select combinations of minimum, most likely, and maximum values. The combination of minimum, most likely, and maximum values needed to construct each distribution type will be conveyed in the following section that describes each type.

SYMMETRICAL DISTRIBUTIONS

The normal (or Gaussian) distribution forms the basis for parametric (classic) statistical techniques such as the Student's t test and others. Because this is a risk assessment book and not a treatise on statistics, and in keeping with my bent for practicality, the complex mathematics of such distributions will not be discussed here. In addition, in my quest to prod the reader into considering distribution form rather than function or equation, I will heretofore refer to such distributions mainly as symmetrical.

Some properties of symmetrical distributions (excluding for now triangular distributions which are described elsewhere in this text) include the following:

- They are bell shaped.
- They are symmetrical about the mean value.
- The mean and median values (and the mode, if there are repeating values as in the case of nominal data) are coincident.

Figure 12.5 depicts a symmetrical distribution. As previously stated, this distribution type forms the basis for many classical statistical techniques. However, I have, except in rare instances, found relatively little use in the world of risk assessment for the symmetrical distribution. This is not to pooh-pooh the statistical prowess and significance of the symmetrical distribution — it is the foundation upon which classical parametric statistical techniques are built. Now, having hopefully placated the statistical police with the preceding statements, I will now comment on the use of symmetrical distributions in practical risk assessment.

When building a distribution, the model builder typically is attempting to represent a data set that they have not actually measured. That is to say that the distribution is a representation of what the actual scatter of values would look like had the person actually taken a statistically significant number of measurements for the variable.

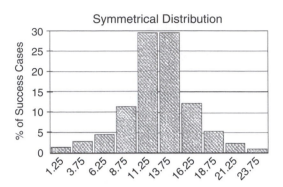

FIGURE 12.5 Symmetrical distribution.

Having performed literally hundreds of risk assessments, I can assert with some confidence that application of a symmetrical distribution rarely is called for.

Experience with many individuals and groups has taught me a few things. Many times, a person performing a risk assessment needs to generate a distribution to represent a variable and does not have a credible notion as to the distribution shape that should be used. In such cases, there is a pervasive tendency for that person to run home to the symmetrical (normal, if you will) distribution type. I am not quite sure why this is so. Perhaps it is because the Gaussian distribution plays such a significant role in classical statistics. Perhaps it is because we typically use the term normal to describe such distributions and the term somehow denotes to the person that it is normal, so it is normally what should be used (imagine that).

In business, science, and other disciplines, those measuring real-world items or events will find that the scatter of actual measurements seldom will be symmetrical. Prices, costs, the size of naturally occurring items (for example, sand-grain size), frequency of occurrence, and other typically measured and ubiquitous risk assessment variables rarely are symmetrical. Even distributions that are used to expand or impart sensitivity to a single value rarely are symmetrical (i.e., multiplying a single-valued-variable coefficient by a distribution whose minimum, most likely, and maximum values are 0.8, 1.0, and 1.1, respectively, for the purpose of causing the single-valued variable to be able to take on values from 20% less than to 10% greater than the single coefficient). When considering ranges around single values, it seems that phrases like plus or minus 10% just roll easily off the tongues of those who really do not have an idea concerning the range. The values greater than a most likely (or modal) value generally are not distributed exactly as those values less than the most likely (or modal) value — a symmetry implied and required by the symmetrical distribution.

Figure 12.5 depicts a symmetrical distribution. Symmetrical distributions built by algorithms I have designed are tailed (tails of the distributions are asymptotic to the X axis) and are created by supplying a minimum and maximum value only (most likely value left blank). The concentration of values around the most likely value (i.e., the peakedness of the distribution) is discussed in a later section of this book.

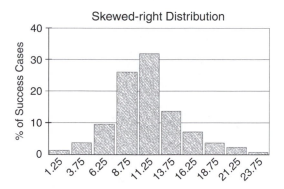

FIGURE 12.6 Skewed-right distribution.

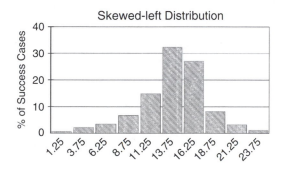

FIGURE 12.7 Skewed-left distribution.

SKEWED DISTRIBUTIONS

The most prevalent distribution type is the skewed distribution. In my practical vernacular, a skewed distribution is defined as one in which the concentration of values (the hump, if you will, in a frequency plot) is not central. The log normal distribution is probably the most popular example of the skewed distribution type. A skewed distribution is shown in Figure 12.6.

A collection of values representing most naturally occurring, business, or technical parameters will be skewed when sorted and plotted. This type of distribution can be skewed to the right or to the left (skewedness is the direction in which the tail goes). Figure 12.7 shows a skewed-left distribution. Skewed distributions built by algorithms I have designed are defined by entering a minimum, most likely (where the hump will be in the frequency plot), and maximum value. The tails of these distributions, as represented by a line connecting the tops of the bars in a frequency plot, are asymptotic to the X axis.

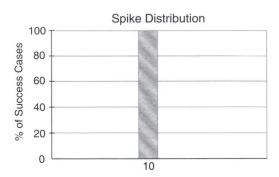

FIGURE 12.8 Spiked distribution.

SPIKED DISTRIBUTIONS

In my terminology, a spiked distribution is defined as a distribution comprised of a single value. Representation of a parameter by a spike often is handy. Figure 12.8 shows such a distribution.

Depending upon the question being asked, a single value can be the optimal representation of the answer to that question. For example, if a risk assessment includes a parameter for the number of people tested, this may in fact be a single value. The number of pages that comprise this book is another example in which a single value, rather than a distribution of values, best answers the question.

Spiked distributions are also useful when attempting to determine whether a given risk assessment model is producing the expected results. In a risk assessment, when input parameters are represented by various types of multivalued distributions and are integrated using Monte Carlo analysis (see Chapter 11), the answer variables will themselves be distributions. With the resulting distributions being a smear of values, it often can be difficult to determine whether the risk assessment model is producing the desired result. In such cases, it often is instructive to represent each of the input variables, for the sake of model testing only, by a spike (single value). Doing so will result in single-valued answer variables from which it is much easier to determine whether the model is producing the expected results. In the distribution-building algorithms I employ, supplying a most likely value only (leaving the minimum and maximum values blank) will produce a single-valued distribution.

FLAT DISTRIBUTIONS

What I refer to as the flat distribution goes by several other names in the books choked with technical jargon. In such volumes, these distributions are most commonly referred to as uniform distributions. Figure 12.9 shows a frequency plot that is the result of 1500 random grabs from a flat distribution.

Unlike the symmetrical or skewed distributions already discussed, the flat distribution has no concentration of values on the X axis. That is, the X-axis values of such a distribution are evenly or uniformly spaced, resulting in a frequency plot (bar

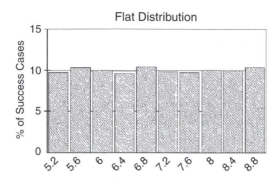

FIGURE 12.9 Flat (uniform) distribution.

chart or histogram) in which the heights of the bars are equal, resulting in a flat (no hump) bar chart. This implies that there is no most-likely value.

The utility of flat distributions is apparent when the user has some idea as to the range of a variable, but has little or no idea what value is most likely. For example, in a given geographic sales area, an area manager may be in the process of opening a new store. For a given product to be sold in the store, the manager knows that in her area the product sells for a minimum of $5.00 per unit (in an economically depressed part of her area) and for a maximum of $9.00 per unit in the upscale areas and in airport and hotel shops. The price for the item is set by what the market in the area will bear. The new store is in a newly developed area for which there is no historical data, so the manager may have no idea what the ultimate price of the product will be. She does know, however, that the price will fall somewhere between the just-above-breakeven price of $5.00 and the maximum price of $9.00. In a marketing risk assessment model, the price for this item may best be represented by a flat (i.e., do not know what the most likely value is) distribution.

The distribution-building algorithms I employ generate a flat distribution when minimum and maximum values are entered (leaving the most likely value blank as with the construction of a symmetrical distribution) and when the user clicks on a toggle button that indicates the distribution is to be flat, rather than symmetrical.

TRUNCATED DISTRIBUTIONS

A truncated distribution is a single-tailed distribution. These distributions can be truncated on the right or left, as shown in Figure 12.10. Truncated distributions are relevant when the minimum value or the maximum value also is the most likely value. For example, electric motors (furnace motors, pool filter motors, etc.) generally run at their maximum or governed number of revolutions per minute (RPM). At times, such as when experiencing abnormally high resistance or when spooling up or down when being turned on or off, the motor may run at less than the maximum RPM. However, running at the maximum RPM is the most likely condition. The RPM situation of such a motor would best be represented by a distribution that is

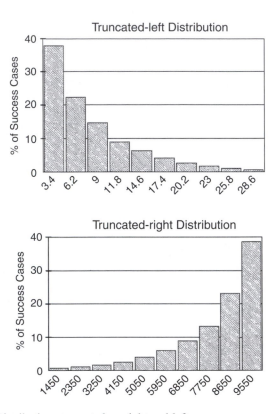

FIGURE 12.10 Distributions truncated on right and left.

truncated on the right — a distribution in which the maximum value also is the most likely value.

In the oil well drilling business, the restart time for offshore oil platforms (rigs) is a good example of a left-truncated distribution. Oil well drilling and producing activities may be halted on an offshore rig for a host of reasons. The approach of bad weather, a safety hazard, broken equipment, inspections, and a variety of other situations can cause a rig to halt operations. For a given rig in a given area there will be a minimum number of days required to restart operations on the rig regardless of how long or short a time operations were interrupted. Transport of inspectors to the rig, safety procedures, mandated start-up processes, and other required steps combine to prescribe a minimum restart time. For a given rig, the minimum restart time may be 2 days — there is no such thing as a restart time of less than 2 days. This restart time also is the typical restart time. That is, most of the instances of rig shutdown require the minimum number of days (2 in this case) to restart the rig. However, occasionally operations on the rig are interrupted because of broken equipment that has to be replaced or damage to the rig by inclement weather. Restart times in these and other cases can range up to 30 days or more. Restart times, therefore, are best represented by a left-truncated distribution in which the minimum and most likely values are coincident (2 days).

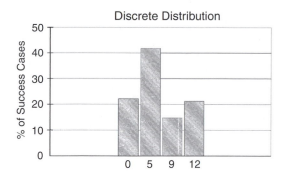

FIGURE 12.11 Discrete distribution.

The distribution algorithms I employ require the user to supply a minimum and a most likely value (with the maximum value left blank) for a right-truncated distribution and a most likely and maximum value only (with the minimum value left blank) to build a distribution truncated on the left. These distributions are tailed in that a line connecting the midpoints of the tops of bars in a frequency plot will be asymptotic to the X axis.

DISCRETE DISTRIBUTIONS

The discrete distribution is the first distribution type described here that cannot be built simply by supplying some combination of minimum, most likely, and maximum values (and clicking on a flat toggle button in the case of the flat distribution). Discrete distributions are of use when attempting to represent a discontinuous variable. For example, the admission price to an event may be $12.00 if you are between the ages of 5 and 65, $9.00 if you are a club member between the ages of 5 and 65, $5.00 if you are a full-time student of any age, and $0.00 if you are over 65 years of age or are 5 years old or under. A risk assessment model may incorporate as an input variable the number of people in each price category who attended the event. Such a distribution is shown in Figure 12.11.

Admission price is a discrete, and not a continuous, distribution. In this distribution there are four discrete values: 0, 5, 9, and 12. When sampling this distribution in Monte Carlo analysis (see Chapter 11), there is no chance of drawing a value of, say, 6 from this set of values. Many real-world parameters used in risk assessment models are best represented by a discrete distribution.

The distribution-building algorithms I use will generate a discrete distribution when the user enters a distribution value (like the 0, 5, 9, and 12 in our example) and a weight for each value. The weights entered are used to determine the relative frequency of the discrete values.

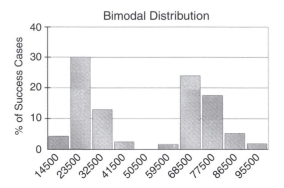

FIGURE 12.12 Bimodal distribution.

BIMODAL DISTRIBUTIONS

A bimodal distribution simply is one that contains more than one most likely value (i.e., more than one hump in a frequency plot, or, in statistical jargon, more than one modal value). Such a distribution is shown in Figure 12.12.

The bimodal distribution depicted in Figure 12.12 is typical in that it appears to be essentially two skewed distributions plotted in the same X–Y space. Such a distribution may be invoked to represent, for example, the return on an investment that depends upon the participation of other key investors. If we are to carry the investment alone, then the return may be in the lower range with a most likely return of around $23,500. However, if we are successful in attracting other investors, our return may be in the higher range with a most likely return of about $68,500.

As previously indicated, a bimodal distribution is essentially two distributions plotted in a common X–Y space. The two distributions need not be skewed as shown in Figure 12.12. Nearly any combination of distribution shapes can be used to create a bimodal distribution. For example, we may have an investment that has the potential to return between $2 million and $5 million if the financial deal does not fall through. If the deal falls through, no return will be realized. There may be a 20% chance that the deal may fall apart, thus, a 20% chance of $0 return. In the bimodal distribution shown in Figure 12.13, there is a 20% chance of drawing (in Monte Carlo sampling — see Chapter 11) a 0 value and an 80% chance of drawing a value of between $2 million and $5 million. This distribution is a combination of spike and skewed distributions.

In yet another scenario, we may be attempting to represent with a bimodal distribution the amount of water processed by a water treatment plant on a given date. The plant may be in the process of being enlarged with the addition of new intake and output pipelines. If the construction is not finished on the date in question, then the plant will run at its old capacity. If, however, the construction of the new pipelines is complete or partially complete on the date in question, then the plant could process water at a rate somewhere between the old processing rate and the new one. However, the most likely scenario is that the plant will process water either at the old rate or at the new rate. In Figure 12.14, two truncated distributions (one

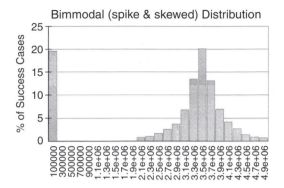

FIGURE 12.13 Bimodal distribution comprised of spike and skewed distributions with associated weights of 0.2 and 0.8, respectively.

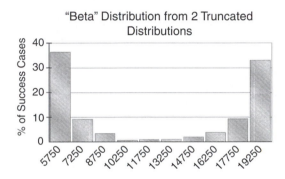

FIGURE 12.14 Simulation of a beta distribution by combining two truncated distributions in a bimodal distribution.

truncated on the left, the other truncated on the right) were used to create a bimodal plot. (Note that this plot is not unlike a beta distribution.)

Most bimodal distribution-building technologies offered by risk assessment packages offer a means of emphasizing one of the distributions (modes) relative to the other. In the algorithms I use, a weight can be assigned to each of the two distributions that compose the bimodal. The weight values simply control the proportion of X-axis values assigned to each of the two distributions that are combined to produce the bimodal distribution. In the case of the distribution shown in Figure 12.13, for example, a weight of 0.2 (20%) was assigned to the 0 (spike) value and a weight of 0.8 (80%) was assigned to the skewed distribution.

READING DATA FROM A FILE

Distributions are used to represent variables because rarely have we measured the entire parent population (i.e., we rarely have made all the measurements it is possible to make). Also, distributions are a convenient method of representing an entire population of values by requiring just a few input values from the person supplying the data (minimum, most likely, and maximum, or mean and standard deviation, or some such set of values).

At times, however, complete or nearly complete sets of data exist that may adequately represent the variable. In such cases, it may be tempting to feed this data directly into the risk model so that the Monte Carlo random-sampling engine can use the actual data as the basis of its sampling. There is nothing technically foul about sampling from raw data rather than from distributions, but there are a few potential problems that might arise.

Distributions built by software systems generally are composed of a discrete number of values. That is, between the minimum and maximum values of a distribution, there are typically 100 to 500 discrete values (this changes with both software package and distribution type) from which Monte Carlo analysis can randomly select. These values usually do not exist as a physical array of values, but are algorithmically generated at run time. Nonetheless, there typically are a limited number of values from which to sample.

One of the factors that dictates the number of Monte Carlo iterations required in a stochastic risk assessment model is the number of values that comprise the distributions. The greater the number of values, the greater the number of iterations required to adequately sample the distributions. (Techniques such as Latin Hypercube can reduce the required number of iterations, but these techniques have problems of their own.) The data values read from a file will be used as the basis for Monte Carlo sampling — just like a distribution. When files containing large numbers of discrete values are employed, it should be obvious that a concomitantly large number of Monte Carlo iterations will be required to adequately sample the array of values in the file. This can have a significant impact on run time and risk program complexity.

Having just warned about the foibles associated with files containing large numbers of values, I now will address some of the problems associated with files that contain a relative dearth of values. It is a relatively rare occurrence that data collected over a finite length of time adequately represent reality. For example, an engineer on an offshore oil drilling platform may have recorded in a file the number of days the platform is shut down (downtime) each time there is a problem of great enough significance to halt operations. The engineer may have been recording data for 5 years. Typical downtimes have been in the range of 2 to 6 days per event. However, in the third year of downtime monitoring, a large 100-year wave (a wave of this size is statistically expected to occur only once in 100 years) struck the platform, destroying critical equipment. The resulting downtime was 30 days. This value of 30 is one of only 10 values that have been recorded over the 5-year period. Monte Carlo random selection will, on average, select the value of 30 one time out of 10. This, effectively, transforms the 100-year wave into the 10-year wave.

Except in the case of short time events, it is a rare situation that we have collected data over a sufficient length of time for our amassed data to represent the real-world situation. Great caution should be exercised when substituting data files for distributions. Make every effort to ensure that a sufficient number of values has been collected so that the range and concentration of the values adequately aligns with the real world.

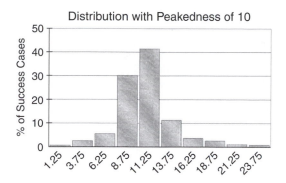

FIGURE 12.15 Distribution with a peakedness of 10.

PEAKEDNESS

From the never-ending battle to keep things simple and practical, the notion of peakedness of a distribution was born. The typical user of a risk model is not versed in the concepts and nuances of statistical analysis, including the concept of kurtosis. Kurtosis can be loosely (but not technically) defined as a measure of the peakedness of a distribution which is, in turn, a reflection of the concentration (or density distribution) of values on the X axis of a frequency plot. Terms such as platykurtic, leptokurtic, and others are used to describe distribution shapes. The kurtosis of a Gausian (normal) distribution is 3.

Risk model users usually do not want to be bothered with this. What they would like to do is simply shape the distribution the way they wish without having to supply mean, standard deviation, kurtosis, or other statistical data. The concept of peakedness is a means by which they can do this.

The peakedness algorithm I employ (which can be replicated in any home-grown risk assessment package) is nothing more than an algorithm that controls the concentration of X-axis values around the most likely value (the hump in the distribution) in those distributions that have most likely values. We explain peakedness to the user as being his or her confidence in the most likely value.

A peakedness range of 0 to 10 has been established. A slider bar is utilized to select a value between 0 and 10. A peakedness of 10 concentrates values near the most likely value, resulting in a frequency plot like that shown in Figure 12.15. In this distribution, the probability of Monte Carlo random sampling selecting a value at or near the most likely value is relatively great compared to the probability of selecting a value in an equally sized range in the tail area of the distribution. By selecting a peakedness value of 10, the user is expressing relatively high confidence that values near the most likely value are those very likely to represent the variable.

A peakedness of 0 puts more values in the tails of the distribution at the expense of the most likely value. Figure 12.16 shows such a distribution. In a 0-peakedness distribution, values near the most likely are still more likely to be selected, but values in the tails of the distribution are relatively likely to be selected compared to a distribution with a peakedness value of greater magnitude. A low peakedness value

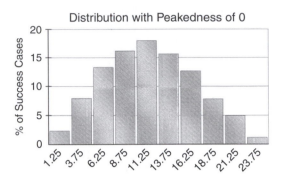

FIGURE 12.16 Distribution with a peakedness of 0.

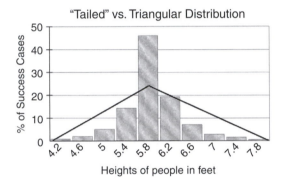

FIGURE 12.17 Tailed distribution contrasted with a triangular distribution.

indicates that the user has relatively low confidence in the most likely value. Using this peakedness scale, the user can express confidence in his or her data and in so doing, control a major aspect of distribution shape.

TRIANGULAR DISTRIBUTIONS

Triangular distributions are ubiquitous in risk assessment models not because they best represent the data, but because they are, in most software packages, easy to build. The typical distribution-building and Monte Carlo-based risk assessment software package requires the user to enter things such as means and standard deviations (and other values that will be covered in the Specific Distribution Types section of this chapter) to build distributions. Most users are, at least, reluctant to supply such data and so turn to the triangular distribution that requires only a minimum, most likely, and maximum value for its construction. This is very attractive.

The fundamental problem with the triangular distribution, however, is that nothing typically incorporated in a risk model is distributed in a triangular fashion. This is more than a cosmetic problem.

Figure 12.17 contrasts a triangular distribution with a tailed (asymptotic) distribution. If we consider the distribution to be comprised of discrete values (like beads on a string) it can be seen in the plot that there is a dearth of discrete data points in a given section of the tail of the asymptotically tailed distribution relative to an equivalent section of the triangular distribution. In the case of the plot in Figure 12.17, in which X-axis values represent heights of adults, the triangular distribution suggests that about one out of two adults is near 7 feet in height (frequency of that height relative to the most likely height). We know that not every second person we meet on the street is of this great height. By contrast, the asymptotically tailed distribution suggests that people of great height are relatively rare.

This discrepancy may seem academic, but the probability exaggeration in the tails of triangular distributions can lead to significant error, especially in the area of portfolio analysis. If the X-axis values in Figure 12.13 were dollars returned on a single investment (instead of height), the combination of multiple distributions in a portfolio-management scheme would lead to significantly inflated upside returns. For this reason and others, triangular distributions should be avoided unless they fit the real-world situation better than any other distribution type.

SPECIFIC DISTRIBUTION TYPES

For lack of a better term, I label as specific distributions all distributions that have been assigned names derived from their authors, their function, the algorithm that generated them, and other sources. Some of the most popular distributions of this category are the chi-squared, beta, gamma, Weibull, exponential, error function, and others.

A common thread of kinship among these distributions is that they require the user to supply distribution-building values that, typically, are difficult for the average person to provide or understand. The beta distribution (effectively two truncated distributions) in Figure 12.18 is an example. Construction of a beta distribution usually requires the user to provide alpha1 and alpha2 parameters (well, is that not nice). A gamma distribution will require you to provide alpha and beta values; a chi-squared distribution needs a nu value; and so it goes.

These distribution types can come in handy. For example, engineers may know from generations of experience that a given parameter is best represented by a chi-

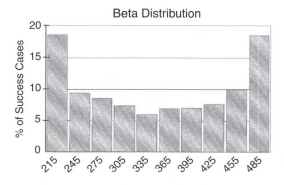

FIGURE 12.18 Beta distribution.

squared distribution. If this is so, then a chi-squared distribution should be used. However, in most situations, the average user will be at a loss to supply the statistical information required to build such distributions, and a simple distribution (as defined previously in this chapter) is a better choice.

SPEAKING DISTRIBUTIONS

A distribution is not simply a collection of specifically arranged numbers for use in a statistical process. As I repeat many times in classes I teach, the best distribution you can build is one you can defend. That is, when asked why a given distribution best represents the situation, just about the last thing the questioner wants to hear as an explanation is some rant about standard deviations or variance or kurtosis.

When one creates a distribution to represent, say, labor costs, defense of the distribution shape should be a compelling story — usually business-based — regarding why a distribution of that particular shape best represents the situation being considered. Let's consider Bob, who owns and runs a business that makes parts for airplanes. Bob has been offered the chance to contribute to a major military airplane-building contract. Bob, however, is apprehensive. Such contracts can be fraught with danger for contributors like himself. The whims of congress regarding funding, powerful congressmen and congresswomen who wish to have certain things done in their states rather than where they can best be done (pork), and huge piles of forms and conflicting regulations are just a few of the pitfalls associated with accepting a role in the big contract.

On the other hand, if it does work out, these contracts tend to be difficult to kill, thereby assuring manufacturers such as Bob's company of long-term income. He decides not to enter into this arrangement before quantitatively examining the situation and considers it prudent to generate some preliminary probabilistic calculations to assess the situation.

Just one of the major considerations in Bob's economic assessment of the situation is the cost of labor. Currently, the average per-hour wage at Bob's company is $10. Bob runs a union shop and the current contract with the union is about to expire. Bob knows that the union will bargain hard for a $12-per-hour wage and the newly-negotiated per-hour rate would be the pay level in effect while his company would be involved in the military contract.

So, what type distribution should Bob use to represent per-hour labor costs in his probabilistic assessment? Consider the high-peakedness truncated distribution shown in Figure 12.19. If Bob used this distribution to represent his view of per-hour labor costs, what would he be trying to say?

The fact that the distribution is truncated at $12 indicates that Bob is not willing to go one penny higher than $12 per hour. This might indicate that Bob knows that his company is the only major employer in the area and workers have few if any employment alternatives. He is willing to go to $12, but if they want one cent more, they know where the door is. The fact that the heights of the bars drop off precipitously from right to left might indicate that Bob has had experience negotiating with the union in the past, and he has found himself lacking in negotiating skill.

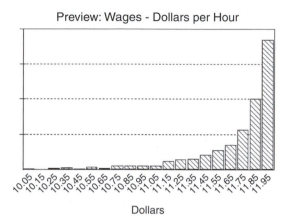

FIGURE 12.19 High-peakedness truncated distribution.

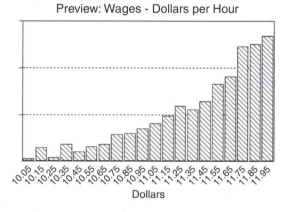

FIGURE 12.20 Lower-peakedness truncated distribution.

The heights of the bars in the $11 range indicate that he believes that he is likely to walk out of the negotiations having agreed to a $12 per-hour rate.

Consider the distribution shown in Figure 12.20. This differs from that shown in the previous figure only in the relative heights of the bars. Such a distribution still indicates that Bob is not willing to go higher than $12, but that he also believes that he has a fighting chance to negotiate a per-hour rate less than $12. Previous experience in successfully negotiating with the union might lead Bob to believe that this distribution shape is the right choice to represent per-hour labor costs in his probabilistic assessment.

Now consider the distribution shown in Figure 12.21. If Bob's business is not the only local employment opportunity for workers, Bob would realize that he is in competition for workers with other employers. Therefore, even though he might not like it, he has to consider that the per-hour wage might be as high as that represented by the high end of this distribution. The tall bars around $12 indicate that he still thinks a per-hour wage around $12 is still most likely in the range and is a very

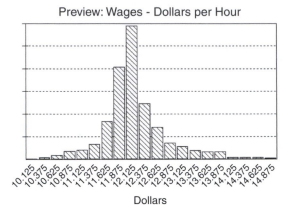

FIGURE 12.21 Per-hour range that strongly suggests a value around $12, but extends to higher values.

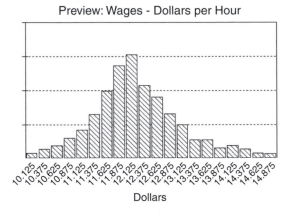

FIGURE 12.22 Per-hour range that indicates that a per-hour wage near $12 is still most likely, but not as overwhelmingly likely as in the distribution shown in the previous figure.

likely outcome of the negotiations, but in his model he has to consider the possibility of higher wages.

This philosophy differs somewhat from one that would cause Bob to use in his model the distribution shown in Figure 12.22. In this distribution, the range is the same as that in the previous distribution — and for the same reasons — but the relative heights of the bars around $12 are reduced. This indicates that Bob still considers a value around $12 most likely, but not as overwhelmingly likely as in the previous situation. Bob might know that alternative employment opportunities are good for workers and although he would desperately like to settle on a per-hour wage near $12, he has to consider that other values are also relatively likely

Considering the two distributions thus far that reached beyond $12, Bob has been viewing that values greater than around $12 are about as likely as those between around $12 and the current wage of $10 per hour. Bob might know from previous

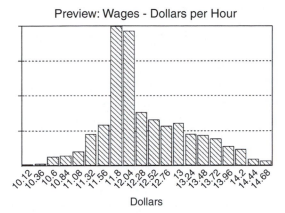

FIGURE 12.23 Distribution indicating that a resulting per-hour wage less than $12 is relatively unlikely compared with one greater than $12.

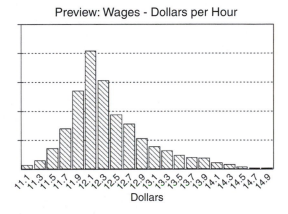

FIGURE 12.24 Distribution that indicates there is no chance of negotiating a value between the current $10 and $11 per hour.

experience, however, that there is a significantly reduced probability of ending up with a value less than $12 than there is of coming out of negotiations with a value in excess of $12. In Figure 12.23, note that the heights of the bars to the left of $12 are reduced relative to those to the right of $12.

Until now, the distributions we have been viewing all supported the philosophy that there was *some* chance — albeit tiny — of leaving the negotiations with a value of $10. This is indicated by the fact that the distributions, on their lower ends, all extend down to a value of $10. Bob might know, however, that there is absolutely no chance that he will leave the negotiating room having talked the union into accepting the current $10 per-hour wage for the new contract. Considering this philosophy, Bob might employ a per-hour wage distribution like that shown in Figure 12.24. This distribution begins at a value of $11, indicating that Bob knows that he has no chance of leaving the negotiating room with a value less than this for the new contract.

So, what is the point here? Do you think that the result from Bob's analysis would be different if he used, say, the distribution shown in Figure 12.19 rather than that shown in Figure 12.24? You bet the results would be different. If someone asked Bob why he used in his analysis, for example, the distribution shown in Figure 12.24, should his explanation center around means and standard deviations and variances and the like? No. The difference between any of these distributions is a difference in philosophy and a person's view of the business, political, and other situations. The argument to support the use of any particular distribution should be the presentation of a cogent business case and a compelling story as to why that distribution was selected. The best distribution you can use is one that you can defend by relating a compelling story.

FLEXIBILITY IN THE DECISION-MAKING PROCESS

It is beyond the scope of this book to address the subjects of option analysis and real options — and I will not try. However, the aforementioned subjects and the decision-making process in business certainly overlap, and I will here attempt to present some of the experience I have gained over the years regarding flexibility in the decision-making process.

In the preceding sections of this chapter, I addressed how to represent uncertainty using distributions. Theoretically, as one progresses through subsequent stages of a project, uncertainty should diminish. I say theoretically, because it can be the case that progression only leads to more choices and uncertainties, but I will ignore that case here.

It is typically the aim of most risk mitigation efforts to reduce uncertainty. In the broadest sense, this is a good thing. However, when it comes to decision making and options, retaining uncertainty can also be beneficial.

I know what you are thinking: "This guy wants it both ways." In a way, I do. As I mentioned at the beginning of this chapter, decisions are necessary when there are choices. If there are choices, then there exists some uncertainty as to which option to take. In business, like in any other part of life, it generally is beneficial to keep your options open. Again, generally, the more realistic the options, the better. So, if we are to have options as we progress through a project, this means having choices. If we have choices, this means that there is uncertainty. Therefore, consciously maintaining some uncertainty is a valid pursuit.

Once one has defined the risks and uncertainties, one should be flexible with regard to changing project focus, be accepting of a range of project results or output, and be flexible in project scheduling and planning. All of this means maintaining options which, in turn, requires maintaining some degree of uncertainty.

For example, consider a business unit leader (BUL) who has to decide which contractor to employ for a given task. Contractor A is the least expensive and exhibits expertise only in the area of the task being considered. Contractor B is more expensive, but has expertise in areas that might be of value later in the project and which might, in fact, be beneficial in execution of the currently considered task.

Even if not selected for the current task, Contractor B is willing to be on call for a price. Contractor B could be called in if Contractor A makes a mess of things, or could be called in at a future date to apply its broader range of expertise to the later parts of the current problem and to future problems. This is a source of income for Contractor B and might be good (and maybe in the end, cheap) insurance for our company.

Keeping this option open — which contractor to use in the future (and possibly now if Contractor A fails) — is not without cost and does bring uncertainty to the process. We have not decided for sure which contractor will be used now and in the future. However, keeping these types of options open and maintaining a certain acceptable level of uncertainty during the execution of a project can be well worth the cost.

The point is, in the end, that the pursuit of eradicating all uncertainty is not always in the best interest of the project. Uncertainty maintained throughout a project's life can be liberating and might just save the day.

SELECTED READINGS

Davis, J. C., *Statistics and Data Analysis in Geology, Volume 2*, John Wiley & Sons, New York, NY, 1986.

Ebdon, D., *Statistics in Geography — A Practical Approach*, Basil Blackwell, Oxford, 1981.

Harbaugh, J. W., Davis, J. C., and Wendebourg, J., *Computing Risk for Oil Prospects: Principles and Programs*, Pergamon Press, Tarrytown, NY, 1995.

Hayslett, H. R., *Statistics Made Simple*, Doubleday, New York, NY, 1968.

Jackson, M. and Staunton, M., *Advanced Modelling in Finance Using Excel and VBA*, John Wiley & Sons, Hoboken, NJ, 2001.

Meyer, M. A. and Booker, J. M., *Eliciting and Analyzing Expert Judgment: A Practical Guide*, Academic Press, London, 1991.

Welch, David A., *Decisions, Decisions: The Art of Decision Making*, Prometheus Books, Amherst, NY, 2001.

Stanley, L. T., *Practical Statistics for Petroleum Engineers*, Petroleum Publishing, Tulsa, OK, 1973.

13 Chance of Failure

CHANCE OF FAILURE — WHAT IS IT?

I will begin this chapter by stating that there is no more common cause for spurious risk assessments and misinterpretation of risk-model results than the exclusion of chance of failure from the analysis. It will be illustrated in this chapter that consideration and integration of chance of failure, when appropriate, is crucial.

As alluded to several times in previous and subsequent sections of this book, the difference between the expected value of success and the expected value for the portfolio is consideration and incorporation of chances of abject failure. More on this later.

Chance of failure (COF), as I write about and define it here, is the probability that the object of the risk assessment will experience abject failure. The total chance of success (TCOS) for the object of a risk assessment is calculated thus:

$$TCOS = (1 - COF \text{ variable } 1) \times (1 - COF \text{ variable } 2)$$

$$\times \ldots (1 - COF \text{ variable } N) \tag{13.1}$$

where N is the number of variables that have an associated COF. Given this definition and calculation of TCOS, it should be clear that the application of the COF concept is valid only in cases in which failure of any variable that has an associated COF will cause the object of the risk assessment to fail. A physical analogy is the proverbial chain made of links that may be thought of, for example, as our project. A project risk assessment is composed of variables. If one link breaks, the chain fails. In the COF scheme, if one variable fails, the project fails. So, the concept of COF, as presented here, is not valid in situations in which failure of one component does not necessarily cause the entire enterprise to fail. A physical analogy is the strands of a rope. Failure of one strand (unlike failure of one link in the chain) does not necessarily cause the rope to fail.

A good real-world example in which COF might be applied is the launching of a new product in the marketplace. We may be considering, among other things, the price we need to get, legislation we need to have passed for our product, and the labor costs. If any one of these variables turns out, in the real world, to be outside our acceptable limits, then the product introduction will fail. That is, if we need a price of at least $12.00 per unit but all the market will bear is $10.00, then the game is over. Likewise, if the government fails to pass the required legislation, we fail. Similarly, we may experience labor costs above what we can bear, causing the enterprise to fail (no matter what other values we experience for other variables).

In contrast, some situations do not warrant the consideration of COF. For example, for our corporation to survive, we may need to attain an after-tax profit of

$10MM. We arrive at our corporate profit by combining the profits from the 17 companies that comprise the corporation. If, in a given year, one or more of the companies fails (i.e., does not realize a profit), this does not necessarily spell failure for the corporation as long as the required profit can be gleaned from the companies that did show a profit.

A good real-world hybrid example is your car. Almost all components of your car have an associated COF. However, if we are building a risk assessment model to calculate our chance of the old car making a 1000-mile trip, consideration of failure for certain components is critical (transmission, alternator, radiator, etc.) while failure of other components (radio, electric locks, etc.) may simply be annoying.

FAILURE OF A RISK COMPONENT

Well, now that I have expounded a great deal about failure of a component or variable in a risk assessment, just how is it that failure is defined for a variable? The risk-model builder has to generate a definition for failure for each variable that has an associated COF. Let us begin with the situation in which COF affects the distribution.

Consider the simple case of the environmental engineer charged with supplying potable water in a drought-stricken area. In Figure 13.1 are shown two distributions: one for sediment load (percentage of a liter of water comprised of suspended solids) and one for size of the underground water reservoir (measured in cubic meters). The percentage of rock particles in a liter of water in the given area has, according to historical data, ranged from a minimum of 3% to a maximum of 20%, with a most likely sediment load of 10%. However, experience has shown that sediment loads greater than 14% are too high, causing the pipes to clog.

The percentage of the histogram bars (frequency) in Figure 13.1 that will fall above 14% is the COF for sediment load. COF generally is expressed as a percentage and is either calculated by the computer algorithm (as in the risk assessment package I use) or estimated by the user. The blackened area in the percent-load plot represents 10.5%, or a COF of 10.5. In this case, failure of the sediment-load variable is defined as the probability that the actual sediment load value will be greater than our maximum acceptable value of 14%.

Now let's turn to reservoir size. The size of reservoirs in the area have ranged from a minimum of 3 million cubic meters (MCM) to a maximum of 90 MCM, with a most likely size of about 35 MCM. Economic calculations indicate that if the reservoir is less than about 20 MCM, then the reservoirs contain too little water to be economically developed. In Figure 13.1, the blackened area in the reservoir size histogram represents 5.5% of the area under the curve and, thus, a COF of 5.5. Failure for the reservoir size variable is defined to be the probability that the actual reservoir size will be less than our acceptable minimum size of 20 MCM.

In a risk assessment incorporating sediment load and reservoir size, the respective input distributions for the two variables would appear as those shown in Figure 13.2. Both of these distributions represent success-only values. Note that it is impor-

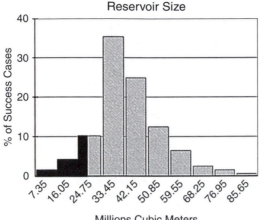

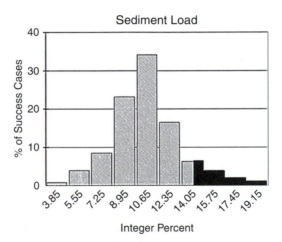

FIGURE 13.1 Frequency plots for sediment load and reservoir size. Black areas represent percent chance of failure (above 14% for sediment load and below 20 MCM for reservoir size).

tant not to allow the probability of the maximum allowable sediment load or the minimum acceptable reservoir size to go to near-zero probability.

The original sediment load and reservoir size distributions have essentially been clipped to produce the distributions shown in Figure 13.2. The parts of each distribution that have been clipped will be used in the calculation of the TCOS of the project thus:

$$TCOS = (1 - 0.15) \times (1 - 0.055)$$

$$= .85 \times 0.945$$

$$= 0.803 \tag{13.2}$$

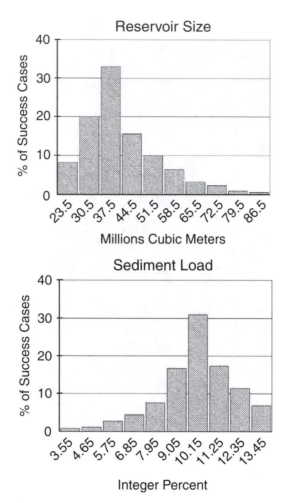

FIGURE 13.2 Sediment load and reservoir size distributions that have been clipped by their respective chances of failure.

CHANCE OF FAILURE THAT DOES NOT AFFECT AN INPUT DISTRIBUTION

In our water project example, one of the input variables is the size of the reservoir. Thus far, this variable has a single associated COF defined to be the chance that the reservoir is of a size that would yield uneconomical volumes of water. However, in the area where our water well is to be drilled, there is some chance that the desired reservoir rock type may have been altered by subsurface hydrothermal fluids (essentially, underground flowing hot water). The fluids may have transformed the reservoir to a rock type that is not of reservoir quality and, therefore, can yield no water at all.

It may be that we have measured the size of the potential reservoir rock just fine (using remote sensing techniques such as seismic or gravity-related technologies), but there is a chance that we are measuring the size of a reservoir of the wrong rock type. Therefore, we may define a second COF associated with reservoir size. This additional COF might be defined as the chance that we are measuring the size of the wrong thing (i.e., the reservoir is composed of a rock type from which we cannot extract water at all). This COF is associated with the reservoir size distribution, but does not clip or in any other way affect the distribution. For the sake of this example, let's say that we believe that there is a 10% chance that we have accurately measured the size of a body of rock of the wrong type.

Yet another variety of COF that does not influence input variable distributions is the type that is not associated with any input distribution at all. Sticking with our water project example, there may exist yet another reason for potential failure of the project. For our project to succeed, we need the local legislature to pass a law that exempts our project from regulations which preclude drilling in our proposed project area. Even though this project is being pursued at the behest of the local population, there is a significant chance that environmental groups may successfully block passage of the required legislation.

In this situation, we may define for our water project risk assessment model an input variable that is a COF only (no associated distribution). The COF-only variable, in this case, might be defined as the chance that we will fail to get the required exemption. Because this variable has no distribution, it will not affect the Monte Carlo-calculated X-axis values but will affect the elevation of X-axis values to form our cumulative frequency plot and the projected X-axis intercept of the curve. Let's assume that we believe there is a 20% chance that the required legislation will fail to be enacted. Given the two additional COF values we have added to our project, the TCOS for the project would now be as follows:

$$TCOS = (1 - COF \text{ sediment load}) \times (1 - COF \text{ reservoir size})$$

$$\times (1 - COF \text{ wrong rock type}) \times (1 - COF \text{ legislation})$$

$$= (1 - 0.15) \times (1 - 0.055) \times (1 - 0.1) \times (1 - 0.2)$$

$$= 0.85 \times 0.945 \times 0.9 \times 0.8 = 0.578 \tag{13.3}$$

INCORPORATING CHANCE OF FAILURE IN A PLOT OF CUMULATIVE FREQUENCY

In the preceding example, our total chance of success for the project is 57.8%. As described in the Output from Monte Carlo Analysis section of Chapter 11 of this book, the usual process for creating a cumulative frequency plot begins with sorting the X-axis values from smallest to largest. The Y axis is subdivided into a number of equally spaced segments (the number of segments equal to the number of Monte Carlo iterations) and the X-axis value of greatest magnitude is elevated one probability unit, the next largest value is elevated two probability units, and so on until the leftmost X-axis value is elevated n (the number of iterations) probability units (Figure 13.3).

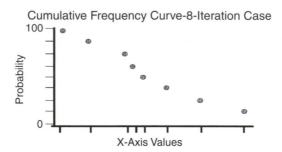

FIGURE 13.3 Individual X-axis values raised parallel to the Y axis according to their cumulative probability.

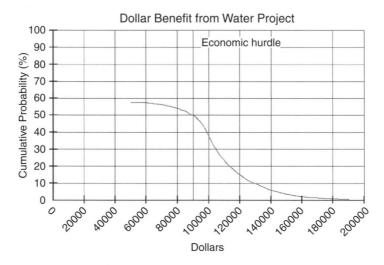

FIGURE 13.4 Cumulative frequency plot that includes chance of failure.

In the immediately preceding section of this chapter, we calculated the TCOS of the water project to be 57.8%. To incorporate COF in the cumulative frequency plot, rather than subdividing the 0–100 Y-axis range into n (number of iterations) segments, the 0-TCOS (in this case, 0–57.8) range is subdivided into n equally spaced segments. The X-axis values are elevated as described previously, resulting in a cumulative frequency plot whose projected left end would intersect the Y axis at the TCOS value (Figure 13.4).

Figure 13.5 shows the same water project data plotted without considering COF. Note that the probability of achieving a dollar benefit amount of $90,000 or greater is about 86%. When considering COF, as shown in Figure 13.4, the probability of achieving the same dollar benefit is only about 50%. This is a significant difference that should result in different decisions being made concerning the water project.

It should be noted that when the cumulative frequency curve is allowed to rise to 100% on the Y axis (no consideration of COF), the probabilities (Y axis) and percentiles (X axis) are correlated. That is, the X-axis value, for example, which

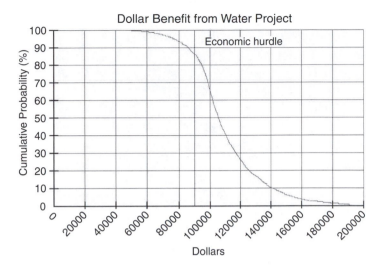

FIGURE 13.5 Cumulative frequency plot equivalent to that in Figure 13.4 but without chance of failure.

represents the 50th percentile (50% of the X-axis values are larger than this value and 50% are smaller) also projects on the Y axis at the 50% probability point. Likewise, the 90th percentile X-axis value (i.e., 10% of the values are larger than this value) projects at 10% probability on the Y axis. This relationship holds true for all percentile/probability pairs. However, when COF is incorporated in the cumulative frequency plot, this percentile/probability relationship no longer applies.

Incorporation of COF does not affect percentiles. For example, the X-axis value that represented the 50th percentile prior to incorporating COF still is the 50th percentile value after COF is considered (COF does not affect the arrangement of X-axis values). However, when the cumulative frequency plot is depressed by COF, the probability of obtaining the 50th percentile point (or a value greater than it) no longer is 50%.

ANOTHER REASON FOR CHANCE OF FAILURE

In the water project example, the cumulative frequency curve was appropriately depressed to account for COF. This is all well and good; however, let's consider the consequence of ignoring COF in a simple risk model in which COF does influence the input variable distributions.

Consider the following simple model:

$$\text{Income} = \$ \, / \, \text{Package} \times \text{Shipment Rate} \, / \, \text{Day} \tag{13.4}$$

in which $/Package and Shipment Rate/Day are distributions with associated COF values. The $/Package distribution may range from $1.00 to $10.00 with a most likely value of $6.00. The associated COF will be defined to be the chance that we will realize a $/Package rate less than the required $4.00. Below this rate, a partner

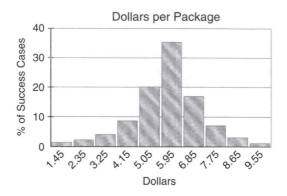

FIGURE 13.6 Frequency plot of $/Package.

FIGURE 13.7 Frequency plot of package shipment rate per day.

company that shares in the $/Package rate will pull out of the enterprise and, consequently, our business will fail. This distribution appears in Figure 13.6.

The independent package shipment rate per day (PSRPD) variable is represented by a distribution with a minimum of 2,000, a maximum of 20,000, and a most likely of 8,000. Our business could survive with a minimum PSRPD of 2,000; however, if the PSRPD gets much greater than 14,000, a significant bottleneck is created at the warehouse, which causes our business to shut down for an unacceptable length of time. So, the associated COF will be defined as the chance that the PSRPD will exceed our warehouse capacity. This distribution appears in Figure 13.7.

If we use $/Package and PSRPD distributions as inputs to a Monte Carlo-based risk model that calculates income from the (admittedly oversimplified) equation

$$\text{Income} = \$ \, / \, \text{Package} \times \text{PSRPD} \tag{13.5}$$

then we could, on these Monte Carlo iterations, draw the following $/Package and PSRPD values:

$$\text{\$ / Package} \times \text{PSRPD} = \text{Income}$$

3	10,000	\$30,000	
5	6,000	\$30,000	
5	15,000	\$30,000	(13.6)

Each of the random selections draws \$/Package and PSRPD values that result in an income of \$30,000. However, two of the three iterations include values that fall within the failure range for one of the variables. Therefore, given our three iteration case, two thirds of the values that would plot at \$30,000 on a histogram of income would not have succeeded in the real world — the histogram would significantly exaggerate our chance of achieving \$30,000 (in this instance, our chance of achieving \$30,000 is only one third of the histogram's indicated probability).

It is tempting to suggest as a solution to this problem that we simply clip off the failure parts of the input distributions, but that is tantamount to hiding our heads in the statistical sand. The values that populate the failure sections of distributions such as these do occur in the real world and have got to be reckoned with. When the concept of COF is employed, the distributions are clipped to exclude the failure values (sometimes the failure area can occur in the middle of the distribution), but are accounted for by depressing the cumulative frequency curve so that probabilities interpreted from the curve reflect the COF.

THE INSERTING ZEROES WORK-AROUND

This thought eventually occurs to nearly everyone who comes to grips with the concept of COF (especially spreadsheet bigots, it seems): "Hey, I could just calculate a zero when I run into a failure situation and I will not have to fool around with this COF stuff." I can tell you from experience that the route of substituting zeros for failure is the road to ruin.

For example, it is tempting to consider inserting some logic in the risk model that will recognize a failure value and so calculate a zero. In our \$/Package and PSRPD example you might write

$$\text{If \$/Package} < 4 \text{ then Income} = 0$$

$$\text{If PSRPD} > 14 \text{ then Income} = 0 \qquad (13.7)$$

and think you have taken care of that nasty failure problem. Well, there are several things wrong with this approach.

First, some risk assessment models are quite complex and can consume a significant amount of time and resources for each iteration. Typically, it takes no less time to generate a zero result than to generate a non-zero result. If the situation includes a relatively high COF, most of the time and effort is spent generating noncontributing zeroes.

In this same vein, if a 1000-iteration risk model includes a 50% COF, this means that 500 zeros will be generated and that the cumulative frequency curve will be composed of only 500 nonfailure values. This would not be so terrible if each situation you needed to run through the model had exactly the same COF. It is the usual situation, however, that individual projects have unique associated COFs. Using the zero-insertion method, you will be left with a curve comprised of 500 points (50% COF) from one run of the model, a second curve made up of 300 points (70% COF) from a second run of the model, and so on. Not only are the curves composed of relatively few values, but the disparity in values can make statistical comparison of curves difficult or invalid.

The clever person might look at this disparity of nonfailure values problem and conclude, "Hey, I know what I'll do, I'll just let the model run until it generates, say, 1000 nonzero values." Nice thought. More iterations generate more zeros. More iterations also increase the chance of sampling and combining extreme values in the tails of the distributions. Under this more iterations scheme, two output cumulative frequency curves generated by combining the same input distributions but having different TCOS values can have significantly different X-axis ranges. This is because more iterations tend to extend the end points of the cumulative frequency curve. This can make it difficult to compare curves from multiple runs of the risk model. A cumulative frequency curve generated using the COF concept will always contain the same number of points as the number of Monte Carlo iterations.

As is my wont, I have saved the best for last. The most glaring problem associated with the substitute zeroes problem is that sometimes zero is a nice benign value that can be used as a failure indicator and sometimes it is not.

Consider the typical risk model. There usually exists up front a technical section followed by, at least, a section that calculates economics (sometimes an engineering section might be sandwiched between the technical and economic parts). In any event, zero might be able to represent failure in one part of a model but would not be interpreted as failure in other sections.

For example, it might be fine to calculate factory output in the first part of a risk model. Factory output is either something (a number greater than zero) or nothing. There likely is no such thing as negative factory output. So, it might be tempting in this case to represent with zero the failure of the factory to produce. An output curve of production, then, might include many zeros indicating the number of times the Monte Carlo process calculated no production.

In a multiple-step risk model, however, the output from one step is passed along to the next. It may well be that zero can represent failure in the calculate production step, but a subsequent engineering step (in which there can be successful negative and positive values) or economics step (in which deficit spending does not represent failure) may not treat zeros as the benign values they were in the production stage of the model. Passing along strings of zeros from one part of a risk model to another is a dangerous practice.

CHANCE OF FAILURE AND MULTIPLE
OUTPUT VARIABLES

A cautionary note concerning the application of the COF concept with multiple output variables. Consider the case in which we are computing two output variables A and B from input variables C, D, E, and F. The equations might be

$$A = C + D$$

$$B = E \times F \tag{13.8}$$

Each of the four input variables may have an associated COF. In this case the risk model should calculate two TCOS values; one for A from the COF values associated with C and D, and one for B from the COF values for E and F. Given the separate and independent equations for A and B, the failure of, for example, E has no affect on A. Likewise, the failure of either C or D will not be considered in calculating the TCOS for B. Great caution should be taken when constructing risk models so that TCOS values for individual output variables are correctly calculated.

A CLEAR EXAMPLE OF THE IMPACT OF
CHANCE OF FAILURE

Let's imagine that we are responsible for estimating the first-year profit expected from our new plant coming on line. Remember, when dealing with chance of failure (COF):

- We define success and, therefore, failure any way we like. For example, success might be defined as our ability to produce a profit of level X in the first year. Success might be defined in terms of time — for example, we will only be successful if we get Y% of the job done by December 31. So, failure is defined to be our inability to achieve whatever criteria we have used to define success.
- COFs are like links in a chain. If any one of the COFs happens, the project is considered to have failed. For example, we might have three COFs associated with a project: (1) the COF that we will not be funded, (2) the COF that we will not get a required permit, and (3) the COF that a change of host government will invalidate our existing contract and render it void. With COF, if any one (or more) of these things happens, the project experiences failure.

Considering our new plant about to come on line, our economists have run their full-cycle economic analysis and have created the first-year-expected-profit plot shown in Figure 13.8.

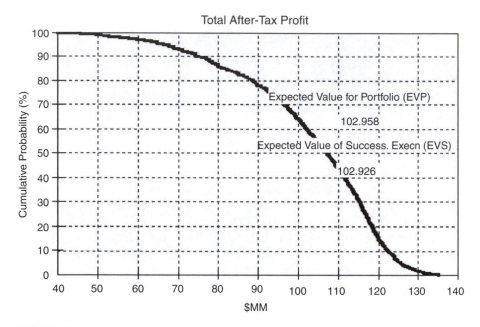

FIGURE 13.8 Cumulative frequency plot of estimated first year profit from new plant without considering COF.

It can be seen from the plot that a successfully executed on-time plant startup should yield an after-tax profit ranging from about $40MM to about $135MM with a mean profit of about $103MM. Note also that, without considering COF, the expected value of successful execution (EVS) and the expected value for the portfolio contribution (EVP) are equivalent.

However, there are some threats to success. We have determined the following:

- There is a 20% chance that we will not obtain a critical permit in time for our plant to start up and produce profit in the timeframe being considered
- There is a 10% chance that we will experience a labor strike that will preclude generating profit from the plant in the timeframe being considered
- There is a 35% chance that critical equipment for the plant will fail to be delivered in time

For the sake of simplicity, we are here using deterministic estimates for the COFs. We can determine the total chance of success (TCOS) thus:

$$TCOS = (1 - Permit\ COF) \times (1 - Strike\ COF) \times (1 - Equip.\ COF) =$$

$$(1 - 0.2) \times (1 - 0.1) \times (1 - 0.35) =$$

$$0.8 \times 0.9 \times 0.65 = 0.468 \times 100 = 46.8\% \qquad (13.9)$$

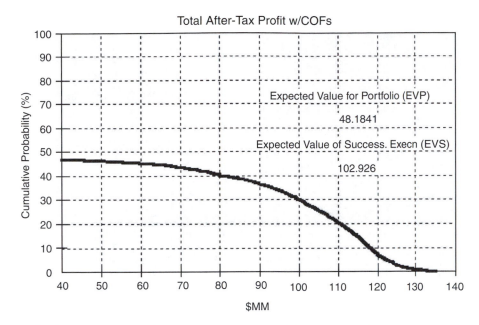

FIGURE 13.9 Cumulative frequency plot of estimated first year profit from new plant considering COF.

In the plot shown in Figure 13.9 it can be seen that the Y-axis cumulative-frequency-curve intercept has been depressed to be at 46.8%. Note that the EVS value is still at about $103MM because this is the value we would experience if none of the COFs were to trigger. The contribution to the portfolio (EVP), however, is now down to about $48MM. Given the COFs, this is the amount the corporation would roll up with other profit projections from other projects to get an estimate of the potential profits for the entire portfolio of yet-to-be-executed projects.

Now, there are those who would adroitly point out that we could have simply multiplied 0.468 by the mean profit of about $103MM to get a value near $48MM (the EVP on the plot). This is true. However, sometimes things get a bit more complicated.

Consider that our economists have indicated that a successful technical startup of the plant could generate the range of first-year profits shown as X-axis values in Figures 13.8 and 13.9. However, technical success will not necessarily bring financial success. The economists have indicated that in order to be able to claim first-year economic success, the plant would have to generate at least $80MM. We can plot this value as an economic hurdle on the cumulative frequency plot as shown in Figure 13.10. Note in Figure 13.10 that there is only about a 40% chance that the first-year plant startup will be economically successful.

The shaded area to the left of $80MM and above the cumulative frequency curve is negative and is added to the shaded area to the right of $80MM and below the cumulative frequency curve to arrive at the EVP value of about $10.7MM. This value is the amount of profit we would roll into a corporate (or other) summation

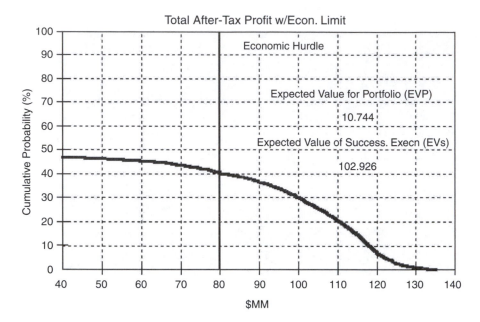

FIGURE 13.10 Cumulative frequency plot of estimated first year profit from new plant considering COF and economic hurdle.

of projected profits from multiple projects. This is significantly different from the about $103MM EVS value that represents the prize we might experience if we can mitigate all of the chances of failure prior to plant startup. The EVP is the amount, considering our current technical and economic situation, the corporation should use for roll ups of similar values from all projects, and to compare this project to any others that might be in competition for support or sanction.

Without the use of COF in the manner demonstrated above, it is tempting to substitute zero for failure. Considering only our technical COF situation, a model might be built that tests each individual-parameter COF against a randomly generated value. If the randomly generated value is, on any iteration of the Monte Carlo process, less than the decimal representation of the individual-parameter COF, then a zero would be used as the value of our answer variable – in this case, first-year profit. Such logic will result in a plot as shown in Figure 13.11.

From this plot, it can be seen that the X axis extends to zero to accommodate the zeros generated by failure situations. In addition, the range of non-zero values will be constrained relative to a process that used the COF technique because there are fewer non-zero iterations that comprise the non-zero part of the curve. The fewer values that comprise the curve, the narrower the range of that curve is likely to be. Separation of the X-axis representation of success from the technical and economic aspects of failure — as when utilizing the COF approach — is most advantageous.

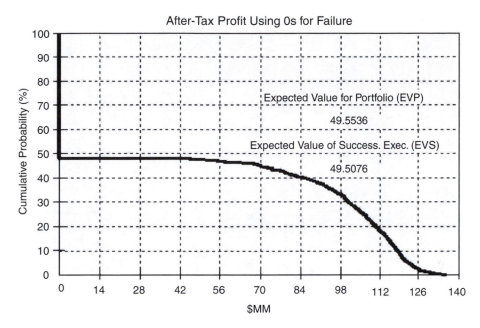

FIGURE 13.11 Cumulative frequency plot resulting from use of 0 as indicator of failure.

SELECTED READING

Nijhuis, H. J. and Baak, A. B., A calibrated prospect appraisal system, *Proceedings of Indonesian Petroleum Association* (IPA 90-222), Nineteenth Annual Convention, 69–83, October, 1990.

14 Time-Series Analysis and Dependence

INTRODUCTION TO TIME-SERIES ANALYSIS AND DEPENDENCE

INTRODUCTION TO TIME-SERIES ANALYSIS AND DEPENDENCE

Chance of failure is a primary risk assessment concept and thus is treated in a separate chapter. There are, however, at least two other data manipulation techniques that are critical elements of real-world risk models. Time-series analysis and dependence are functions that help risk models better emulate actual situations.

Time-series analysis allows us to break free of the single period assessment and to project the analysis through time. This is a facility that is critical in most business applications, for example, in which cash flows over time are combined to generate project-representative values such as net present value (NPV).

Dependence is a technique that allows a risk model to respect the relationships between variables. Without application of dependence technology, any hope of emulating real-world processes is folly. Both time-series analysis and dependence are treated in the following sections.

TIME-SERIES ANALYSIS — WHY?

The desired outcome of a risk analysis is, of course, a stochastic assessment of the situation at hand. In a business setting, regardless of the technical, environmental, political, or other nature of the risk model, ultimate decisions are primarily based on a probabilistic financial assessment of the opportunity. Financial measures such as discounted return on investment (DROI), internal rate of return (IRR), and NPV are typical measures upon which decisions are based. Financial considerations of this ilk require the evaluation of an opportunity over time, commonly some number of years. An NPV, for example, is based upon a series of cash flows.

It is beyond the scope of this book to explain the nuances of the calculation of measures such as NPV, the definition of which can be found in nearly any basic financial text (see selected readings for this chapter). It is, however, precisely the premise of this chapter to explain how a risk model can result in the calculation of yearly cash flows and financial measures such as NPV.

TIME-SERIES ANALYISIS — HOW?

Economists working for most companies these days typically employ spreadsheets of one type or another to handle the financial attributes of a project. It is quite common to see in such spreadsheets some number of columns designated as years

Year	1998	1999	2000	2001	2002
Labor costs (in thousands $)	900	600	550	500	500
Capital costs (in thousands $)	1000	500	400	400	400
Yearly production (tons)	30000	35000	40000	45000	45000

FIGURE 14.1 Spreadsheet with time periods as columns and variables as rows.

with rows below the yearly column headings representing variables in an analysis. Variables such as yearly labor costs, capital costs, yearly production, and other parameters are typical (see Figure 14.1). In a deterministic spreadsheet, these parameters and others would be combined to calculate, for example, a single NPV. In the risk assessment world, however, we strive for a distribution of answer-variable coefficients.

Generating a distribution for an answer variable such as NPV is accomplished by combining input variables that are themselves distributions. Using spreadsheets and a spreadsheet add-on risk assessment package, the yearly and single-valued parameters (i.e., labor costs, etc.) can have each of their yearly values expanded into a distribution. That is, rather than having the 1998 labor costs in Figure 14.1 represented by the single value of 900, 1998 labor costs might range from a minimum of 700, a most likely of 900, and a maximum of 1000.

Several techniques may be employed to transform single values into distributions. For example, it may be specifically known that 1998 labor costs will range from 700 to 1000 and that 1999 labor costs will range from 500 to 750 and so on for subsequent years. In the risk assessment software, then, a separate distribution can be established for each year for each variable. This technique has the advantage of allowing the risk model user to specifically control the parameter values through time. Several disadvantages are inherent in this methodology. First, it is rare, indeed, that we in fact know specific ranges for years beyond the current year. Also, it should be clear that implementing this type of single value expansion puts the onus on the user to build an inordinate number of distributions (one per year per variable). Generally, this is impractical.

Another method commonly employed to expand single values into distributions is to establish one or more expansion distributions. For example, we may believe that our labor costs in the first 3 years of the project will range from 10% less to 20% greater than the single value in the spreadsheet. We can then, for the first 3 years, multiply the single spreadsheet value by a distribution that ranges from 0.9 to 1.2. If we believe that for some number of subsequent years different ranges are valid, other expansion distributions can be established and applied in the appropriate time frames. This method of single value expansion reduces the year-to-year control of the parameter distribution, but also reduces the number of distributions the user is required to supply. On each iteration of the Monte Carlo process, the 0.9 to 1.2 distribution will be sampled and the randomly selected value will be multiplied by the single value in the spreadsheet.

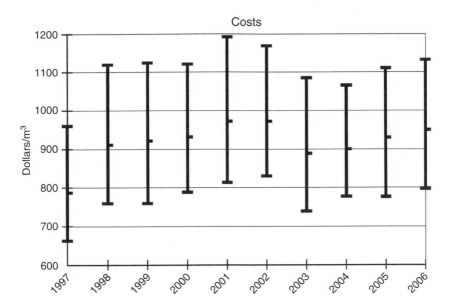

FIGURE 14.2 Typical time-series plot.

Thus far, the time-series examples in this chapter have involved the combination of a risk modeling package and a spreadsheet. However, any complete and self-respecting risk assessment software package should have the self-contained where-withal to effect time-series analysis without need of a spreadsheet. Such risk packages must contain a means by which any number of time series can be established and a syntax for manipulating multivariate analyses over time. See Chapter 17 of this book for an illustration of such an analysis.

TIME-SERIES ANALYSIS — RESULTS

A risk analysis equation generally terminates in the calculation of a resultant variable that is represented by a distribution. Because a time-series analysis generally considers a proposition over a number of years (i.e., a distribution of years), a resulting time-series answer-variable plot must be able to depict a distribution of distributions. Such a plot is shown in Figure 14.2.

In Figure 14.2, each vertical bar represents the distribution of values for 1 year in a time series. The short horizontal bars at the bottom and top of each vertical bar represent the minimum and maximum values, respectively, for that time period. The third horizontal bar represents either the mean or median value of the distribution. Positioning the computer's cursor on any vertical bar and clicking on it can result in the display of the distribution for that time period (Figure 14.3). In this example, a time series of cash flows was generated. From such a time series, an NPV distribution for the project can be calculated.

In reality, a time series plot such as that in Figure 14.2 knows nothing about time. X-axis increments can represent any set of time periods such as months, days,

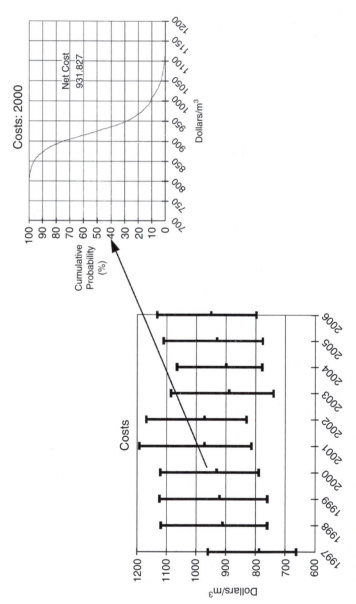

FIGURE 14.3 Time-series plot with expanded single period.

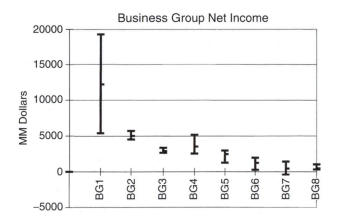

FIGURE 14.4 Time-series plot in which X-axis entities are not time periods.

or hours. In fact, the X-axis intervals need not represent time at all. For example, each tick mark on the X axis might represent a different investment opportunity with each vertical bar indicating the range of return for that investment. An example is given in the next section.

TIME-SERIES ANALYSIS THAT IS NOT

A conventional time-series plot is just that — a plot of a series of time periods. The bars on the plot represent the range of answer-variable values for a given time period. A time-series plot itself, however, actually has no predisposition toward delineating time on the X axis. That is, the tick marks along the X axis need not represent segmented time.

For example, consider the corporation with seven operating business units. Corporate management might be interested in comparing the net incomes from each unit for a given year. A time-series plot can be constructed in which the X-axis tick marks represent the names of business groups while the Y axis, in this case, represents net income in dollars. Such a plot appears in Figure 14.4. Similarly, a law firm might wish to compare strategies concerning a potential litigation situation. In this case, X-axis tick marks could represent options such as settle, arbitration, adjudication, or jury trial.

SOME THINGS TO CONSIDER

Time-series analysis is a powerful analytical technique and an oft-employed process in Monte Carlo-based risk models. The marriage, however, can result in some algorithmic behaviors of which you should be aware. In traditional Monte Carlo analysis, each distribution representing a risk-model variable is sampled once on each iteration. Classic Monte Carlo analysis would, on each iteration, combine the selected values according to the equations to arrive at an iteration-specific answer. The introduction of time-series analysis complicates this simple scheme. If, for example, we decide to calculate NPV based on a 5-year projection of cash flows, then on each iteration the set of equations would be executed five times — once for each year.

On the face of it, this seems reasonable and seems to introduce only a slight wrinkle in the fabric of traditional Monte Carlo processes. However, consider the following situation. We have projected margins for the next 5 years. We expect our margin to grow over the years by 10 to 20%. This growth is represented by a growth distribution that ranges from 0.1 to 0.2. Multiplication of the year's margin value by a randomly selected value from the growth distribution will cause our margin to increase by between 10 to 20% — just what we wanted. Or is it?

Recall that each distribution is sampled once on each iteration. Remember also that on each iteration when we employ time-series analysis, we apply the randomly selected values in each period of the time series — in this case, in each of 5 years. So, if on the first iteration we select a growth rate of 15%, we will grow our margins at 15% every year. Is this what you meant and is this what happens in the real world?

Real-world yearly growth is somewhat erratic. That is, margins may grow by 12% this year, by 17% next year, by 10% the following year, and so on. If we want our risk model to emulate the real world, then we need to employ a mechanism that allows the sampling of distributions not only by iteration, but by time-series period. The risk assessment software I employ facilitates this type of period-based sampling and analysis. You should also select a software package that accommodates this process.

Well, so what if I cannot or do not resample the growth distribution on each iteration? What possible difference could it make? As it turns out, a big one.

Consider the situation in which we do not resample the growth distribution across periods. In a many-iteration model (it is not uncommon to run such models for thousands of iterations), it is statistically safe to say that in a distribution that ranges from 0.1 to 0.2, on at least one iteration a value very near 0.2 will likely be selected. Likewise, it is likely that a value at or near 0.1 will be selected on another iteration. Because these values will be applied in each year, the resulting year-by-year plot of margins will have a much greater range (for each year) than would a plot resulting from period-by-period sampling of growth. In the single-selection case in which 0.2 was selected as the growth rate, each year will grow by this amount. This will generate a significantly higher margin upper bound than if the growth rate were resampled for each period. The value of 0.2 might be selected for year one, but lesser values are likely to be selected for subsequent years. This will significantly lower the upper bound for margin projections. Likewise, selection of 0.1 and employment of that value across all periods will give an unrealistically pessimistic view of margin growth.

Well, having been convinced that this is a real and acute problem, we have purchased or built a risk system that can accommodate period-based sampling of distributions. So now we've got it licked, right? Wrong. There is a little thing called dependence that has to be considered.

Traditionally, when dependence between two variables is established, one variable is designated the independent variable and the other the dependent variable. This Monte Carlo process first randomly samples the independent variable. Then, depending on the degree of correlation between the two variables, the process establishes a dependent variable value that is reasonable for the independent variable coefficient. This process happens once for each iteration.

Well, now we are resampling our distributions not only for each iteration, but across *n* periods within an iteration. If the distribution being resampled period by period has dependence relationships with the other variables, then those relationships need to be honored across periods, not just across iterations. Again, a risk-model-building software package should be purchased or built that will accommodate such a process.

DEPENDENCE — WHAT IS IT?

In the real world, parameters that are co-contributors to a risk assessment may be independent. For example, as part of a total income calculation a risk model might contain two distributions for the variables Number of States in Which We Will Operate in the Year 2000 and Number Of Countries in Which We Will Operate in the Year 2000. In this business, the domestic and international arms of the corporation are separately managed and funded. In the risk model, then, the aforementioned distributions can be independently sampled and no invalid pairs of values can be generated (i.e., in the risk model, any number of states can be combined with any number of foreign countries).

In an actual risk model, many variables exhibit dependent relationships. For example, in the risk model we use in the domestic part of the corporation, our income calculation considers variables such as Stage of Technology and Customer Commitment. For both variables, we might generate a distribution. In the risk model, we would not want to independently sample these two distributions because such sampling might combine, in a single Monte Carlo iteration, high Stage of Technology values with unrealistically low values for Customer Commitment. These two variables have a relationship, in that as our corporation is increasingly viewed as the leader in our technical field, the commitment of customers (percentage of customer base, number of long-term and binding contracts, etc.) increases. The concept of dependence is that of honoring the relationship between two or more variables.

INDEPENDENT AND DEPENDENT VARIABLES

To invoke the concept and underlying technology of dependence, the first item on the agenda is to decide, for a pair of variables exhibiting dependence, which is the independent and which is the dependent variable. That is, which variable controls the other? .

In the case of the Stage of Technology and Customer Commitment variables, it may be clear in our corporation that our customer commitment increases as a result of our technical prowess (as opposed to our technical ability increasing because we are exposed to more and varied customers). In a different company, therefore, it might be that increased technical ability results from the diversity and size of the customer base.

In our situation, Stage of Technology would be considered the independent variable and its distribution would be randomly sampled by the Monte Carlo process. The value selected from the Customer Commitment distribution would, on each iteration, depend upon the Stage of Technology value selected. Thus, Customer Commitment is the dependent variable.

For cases in which it cannot be easily discerned which is the independent (controlling) and which the dependent (controlled) variable, the independent variable should be selected based upon your confidence in a given variable. For example, in another company, it may not be clear which of our two variables (Stage of Technology or Customer Commitment) is the controlling entity. They may know, however, that the values in our Customer Commitment distribution were gleaned from a comprehensive, formatted, and rigorous survey of all customers. Values that comprise the distribution for Stage of Technology were generated in a meeting of managers, many of whom were not intimately familiar with the technology. In this case, they will have much more confidence that the Customer Commitment values actually represent reality and, therefore, the Customer Commitment variable would be selected as the independent variable.

DEGREE OF DEPENDENCE

When it has been established in a risk assessment that we indeed have a dependent relationship between two or more variables, and when we have decided which of the variables will serve as the independent variable and which as the dependent parameter(s), the next hurdle is to establish the strength of the relationship (dependence) between the variables. The strength of a relationship can be calculated or estimated in several ways.

The strength of a dependence usually is expressed as the degree of correlation between two variables (see selected readings in this chapter for references relating to correlation). Correlation can loosely be described as the degree to which one variable tracks another. Correlation is usually expressed on a scale of –1 to +1, with 0 indicating no correlation at all (no relationship between the variables).

Figure 14.5 is a graphical representation of two variables that are independent; i.e., they are uncorrelated and have a correlation coefficient of 0. Two such variables might be the number of eggs used on a day of business in the corporate cafeteria and the number of cars that passed by the building that day. Because the number of eggs used and the number of cars passing by have nothing to do with one another, for any number of cars (any X-axis value), any number of eggs (any Y-axis value) can be chosen. So, knowing how many cars passed by on a given day does not help at all in predicting how many eggs were used on that same day. There is no correlation and, therefore, a correlation coefficient of 0 is appropriate.

The opposite end of the correlation spectrum is perfect correlation, indicated by a correlation coefficient of –1 or +1. A –1 value indicates perfect negative correlation and +1 a perfect positive relationship. Well, what does that mean?

The minus (–) part of the –1 correlation coefficient means that as the value of one variable increases, the value of the other variable decreases. The 1 means that the correlation is without variance. A good real-world example is the gumball

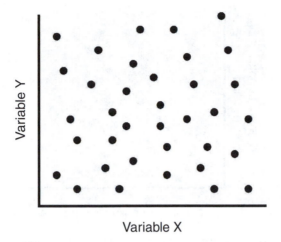

FIGURE 14.5 Cross plot of two independent (uncorrelated) variables.

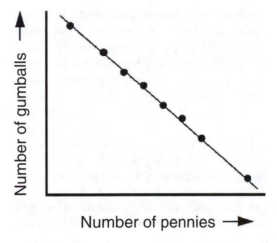

FIGURE 14.6 Cross plot of two variables with a correlation of near –1.

machine in the company lounge. Concerning the gumball machine, we may record how many gumballs are in the machine each day (assuming it is not refilled) and how many pennies are in the machine (I will here give some indication of my age by assuming that each gumball costs 1 penny — oh, the good old days.). After recording the number of gumballs and the number of pennies each day for 6 months, we plot the data and find that it looks like the plot in Figure 14.6.

From this plot, we find that if we know the number of pennies, we can predict exactly (a correlation of 1) the number of gumballs left. We also know that this is a negative relationship (the – part of –1), because as the number of pennies goes up, the number of gumballs goes down.

In Figure 14.7 is shown a plot of perfect (1) positive (+) correlation. This is a plot from corporate security which indicates what they wish were true. The plot

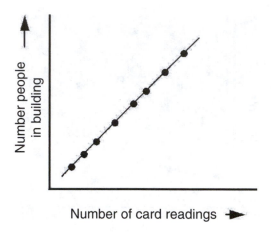

FIGURE 14.7 Cross plot of two variables with a correlation coefficient of near +1.

shows the difference between the number of access-card readings from people entering the building less the number of cards read when people left the building plotted against the number of people in the building. According to security theory, the number of cards read for people coming in less the number of cards read for people exiting should exactly match the number of individuals in the building at any given time. So, knowing the difference in card readings (X-axis values) should exactly predict the number of people (Y-axis values). This should be a perfect (1) positive (as values for one variable increase, values for the other variable increase) relationship.

The fire marshal has been to visit us and has noted that we have only two fire doors. She believes that we have more people in the building than only two doors can handle and, thus, thinks we need at least a third door. Corporate security has been charged with counting noses.

Security thought that they could tell exactly how many people were in the building by examining their card-reading data. However, several counts of bodies revealed that the actual number of people in the building was not exactly predicted by the card reading data. They discovered that it is typical, on both entering and leaving the building, that one person will open the door with their card and several people will enter or exit while the door is open.

Therefore, while it is true that the number of card readings is some indication of the number of people in the building, it is not the perfect predictor (+1) they would like. So, if the number of cards read does give them some indication of body count, the correlation between the two variables is not 0. Likewise, it is not +1 because it is not a perfect predictor. It follows, then, that the relationship between these two variables should be indicated by a value somewhere between 0 and +1.

If historical data are available (as in the case where corporate security has card reading and actual body count data), the typical method employed to establish a relationship strength is to load the data into a software package that will perform a linear (or nonlinear if you want to get really fancy) regression. Linear regression models will return an r-squared value (correlation coefficient) that will give a rea-

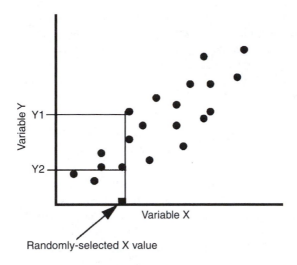

FIGURE 14.8 Cross plot of two variables with a less-than-perfect correlation. Range Y1–Y2 represents range of reasonable Y values for the randomly selected X value.

sonable estimate (for linearly varying data) of the strength of the correlation or relationship between the two variables. That is, how good a predictor of the dependent variable (body count) is the independent variable (card readings)?

Again, 0 indicates that it is no good at all; 1 indicates that it is perfect. Neither of these is true in this case. Therefore, a plot of card readings vs. number of bodies would look something like the plot in Figure 14.8. A regression model run on these data might return a correlation coefficient of, say, 0.8. This indicates that for any value of card readings selected, there is some reasonable range (Y1 to Y2 in Figure 14.8) of Y-axis values (number of people in the building).

Remember, a correlation of 1 yields a single point on the Y axis for any X-axis value selected. A correlation of 0 means that for any X-axis value selected, any Y-axis value is valid. So as we move from 1 to 0, the range of Y-axis values that would be reasonable matches for a given X-axis value gets greater and greater until, at 0, the entire range of Y-axis values is available.

So if we determine that there is a dependence between two variables, and if we then decide which variable is independent and which dependent, and if we assign, based on historical data and regression analysis, a correlation coefficient to the relationship, then the dependent variable will track the independent variable. That is, on each iteration of Monte Carlo analysis, the independent variable will be randomly sampled, and the value for the dependent variable will be randomly selected from a restricted range (like that of Y1 and Y2 in Figure 14.8) within its distribution.

When historical data are not available and when you know that two variables are correlated, an educated guess will have to be taken as to the magnitude of the correlation coefficient. This is not as bad as it may sound. First, it is most important that two correlated variables are made to track one another. Often, the exact (if there is such a thing) calculation of a correlation coefficient is less important. In my classes

I generally have students first assign a strong, moderate, or weak label to each variable pair that exhibits some correlation. A weak relationship can be represented by a correlation coefficient between something greater than 0 and ±0.55, a moderate relationship by a coefficient between ±0.55 and ±0.75, and a strong relationship by a coefficient greater than 0.75. This certainly is not an exact science, but it generally serves the purpose.

MULTIPLE DEPENDENCIES AND CIRCULAR DEPENDENCE

In any given risk assessment there can be many pairs of variables that exhibit dependence. For example, the number of employees might correlate with the number of sick days taken (positive relationship — as one goes up, the other goes up) and the net income for our business unit may correlate with our costs/expenses (negative relationship — as costs go up, income goes down). It may also be true that one independent variable can have several associated dependent variables. Going back to our Stage of Technology example, it may be that as an independent variable, Stage of Technology may control several dependent variables. We already discussed the fact that at our corporation, as Stage of Technology increases, our Customer Commitment increases. This, then, is a positive correlation and would be indicated by a correlation coefficient between 0 and 1. However, it may also be true that our Stage of Technology also affects our corporation's cash exposure. That is, as we are increasingly viewed on Wall Street as the leader in this technology, it is increasingly easy for us to attract investors, thus lowering our corporation's cash investment. So, as Stage of Technology increases, corporate cash exposure (investment) decreases. This is a negative relationship that would be represented by a correlation coefficient between 0 and –1.

We can set up as many variables to be dependent on a single variable as we like (i.e., as many children dependent on a parent as we like). In this scenario, though, one must be careful not to set up circular dependence. Any software package worth the CD it comes on should automatically guard against circular dependence. For example, for three variables, A, B, and C, circular dependence is employed if I say A is dependent on B, and B is dependent on C, and C is dependent on A. This circular logic should be rejected by the system. In this simple example, the circular reasoning is obvious. However, in complex technical and economic risk assessment models, it can be easy to inadvertently attempt to invoke such relationships.

EFFECT OF DEPENDENCE ON MONTE CARLO OUTPUT

Consider two independent (correlation of 0) distributions A and B from which we will calculate C by the formula C = A + B. The distribution for variable A ranges from 2 to 10 with a most likely value of 4. Distribution B ranges from 20 to 30 with a most likely value of 24. Both distributions are shown in Figure 14.9.

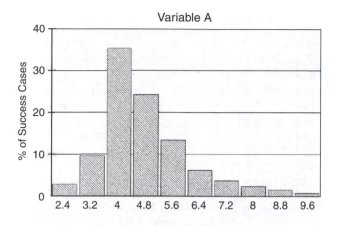

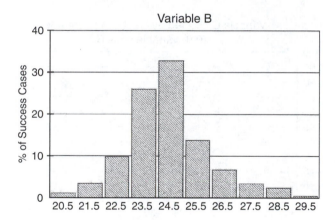

FIGURE 14.9 Frequency plots of independent variables A and B.

On a given Monte Carlo iteration, if an extreme value of A is selected (say, a value of 9), it most likely will be combined with a near-most-likely value from B. The resulting cumulative frequency plot for C would look like that in Figure 14.10 and might range from a minimum C value of 23.7 to a maximum C value of 36.4.

If, however, a strong dependence of, say, 0.9 is established between A and B, then on any given Monte Carlo iteration, when an extreme value of A is selected, dependence will force an extreme value of B to be combined with that of A. This will result in the plot of C shown in Figure 14.11, in which C now ranges from 22.4 to 40.0. A major effect, therefore, of implementing dependence is to change (generally expand) the range of the answer variable.

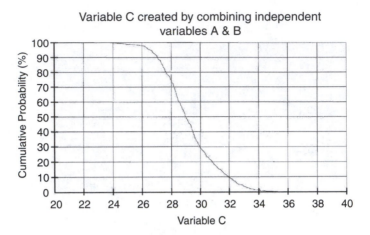

FIGURE 14.10 Cumulative frequency plot resulting from using Monte Carlo analysis to combine two independent variables A and B.

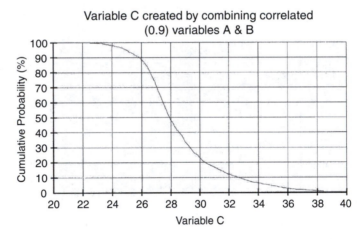

FIGURE 14.11 Cumulative frequency plot resulting from using Monte Carlo analysis to combine two highly correlated (0.9) variables A and B.

DEPENDENCE — IT IS UBIQUITOUS

Whenever I offer a risk class for managers, the class is given sans computers (some managers may read this, so I will not go into why we do not use computers). In a computerless environment, exercises must be designed to bring home a point without venturing too far into the erudite technical haze. One of my favorite computer-free exercises is one I use to illustrate the ubiquitous nature of dependence and the need to consider it.

In Figure 14.12 is a list of variables from an actual plant-construction model. The original list was much longer. The list presented here has been whittled down for simplicity.

Dependence

Public perceptions
Market research
Cost analysis
Production capacity
Demand
Profit margin per unit
Conversion costs (depreciated)
Conversion costs (actual)
Advertising costs
Litigation costs
Operating revenue
Tax rate
Marketing synergies
Taxable income

FIGURE 14.12 List of variables in dependence exercise.

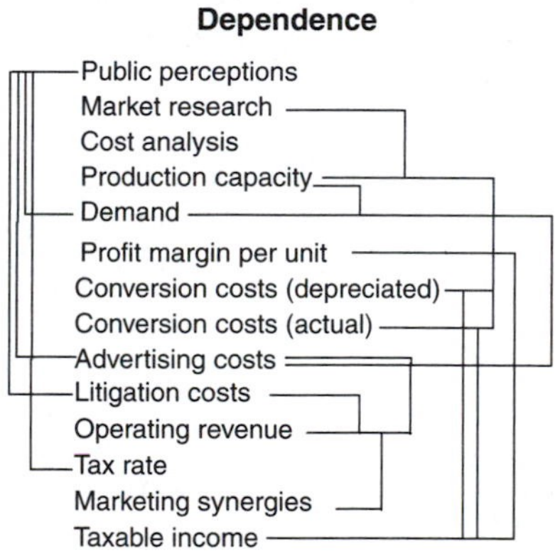

Dependence

FIGURE 14.13 List of variables with dependencies shown as connecting lines.

I typically challenge the class to peruse the list and then discuss which pairs of items on the list have real-world relationships (i.e., dependencies). I always get them kick-started by offering an example. The suggestion is put forth that public perceptions and litigation costs are linked, the reasoning being that if you are perceived as an evil empire by the public, you will spend more time in court. I then draw a line connecting these two items indicating that they have a relationship. As public perception of your company worsens, litigation costs go up. I then challenge the group to come up with some connections of their own. Typically, some lively discussion ensues.

The number of lines connecting variables in Figure 14.13 is about half of what can be drawn (I excluded the rest of the lines because it just makes too big a mess). As can be seen, there are many relationships, that is, dependent-variable pairs. In class, I project the original list on a white board or flipchart on which I draw the

Dependence

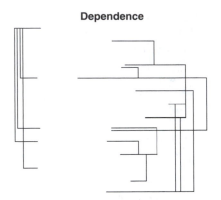

FIGURE 14.14 Lines indicating dependencies.

connecting lines. When the projector is turned off, you are left with just the tangle of lines shown in Figure 14.14 (remember, only about half the lines are shown in Figure 14.14). Class members generally are impressed concerning the number of relationships and by the fact that there is nary a variable on the list that can be independently sampled — that is, sampled without consideration of at least one other variable.

I make the point that if you wish your risk model to emulate the real world, then the relationships shown must be honored. Even using such a simple list of variables, the relationships can be many and complex. If not considered, your risk model will be emulating something, but likely not what you intended.

SELECTED READINGS

Brockwell, P. J. and Davis, R. A., *Introduction to Time Series Analysis, 2nd Edition*, Springer-Verlag, New York, NY, 2002.

Campbell, J. M., *Analyzing and Managing Risky Investments, 1st Edition*, John M. Campbell, Norman, OK, 2001.

Campbell, J. M., Jr., Campbell, J. M., Sr., and Campbell, R. A., *Analysis and Management of Petroleum Investments: Risk, Taxes, and Time*, CPS, Norman, OK, 1987.

Chatfield, C., *The Analysis of Time Series, 6th Edition*, Chapman & Hall/CRC Press, Boca Raton, FL, 2003.

Davis, J. C., *Statistics and Data Analysis in Geology, Volume 2*, John Wiley & Sons, New York, NY, 1986.

Koller, G. R., *Risk Modeling for Determining Value and Decision Making*, Chapman & Hall/CRC Press, Boca Raton, FL, 2000.

Van Horne, J. C., *Financial Management and Policy*, Prentice Hall, Englewood Cliffs, NJ, 1974.

15 Risk-Weighted Values and Sensitivity Analysis

INTRODUCTION TO RISK-WEIGHTED VALUES AND SENSITIVITY ANALYSIS

When a risk assessment is completed, interpretation of the results is the next step. Making this task as effortless as possible can be critical to the success of a risk model. If output from a risk assessment cannot be understood by those to whom it is presented, it likely was a wasted effort to have pursued the risk analysis in the first place. Risk-weighted values facilitate the integration of several risk-model output components. The various types of risk-weighted values make simple the interpretation and comparison of output from multiple assessments.

Likewise, sensitivity analysis aids in identifying the elements of a risk model that were most and least important to the calculation of the answer. Following a risk-model run, examination of a sensitivity plot can be a fundamental step in deciding what action to take as a result of the analysis.

RISK-WEIGHTED VALUES — WHY?

The reason we need risk-weighted values always reminds me of a dog chasing a car. The dog races down the street, expending copious quantities of energy, and finally catches the car. The question now becomes, "Okay, buddy, you caught it — now what are you going to do with it?"

So it is with risk assessments and their output. We struggle arduously to generate a detailed, comprehensive, and practical risk model and its resulting plots and erudite statistics. Well, okay, buddy, you got the results, now how are you going to interpret and compare output from multiple risk model runs?

There is use in assessing the magnitude of benefit or cost and the related probabilities associated with a single and isolated event. Relatively large returns, however, only are realized when we undertake the comparison of multiple scenarios or events. Without employing risk-weighted values, the process of comparison of risk-model outputs can be difficult and at times practically impossible.

Consider two investment opportunities, A and B. Opportunity A has a 20% chance of falling through. That is, if we invest in opportunity A, we have a 20% chance of losing our investment and realizing no return. If opportunity A does not fail, it will return no less than $3 million, no more than $12 million, and will most likely return about $5.3 million.

Like investment A, investment B has a chance of failure. There is a 35% chance that opportunity B will fail and our investment will be lost. If investment B does

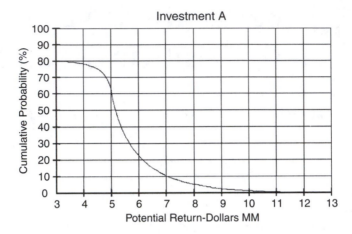

FIGURE 15.1 Cumulative frequency plot for Investment A.

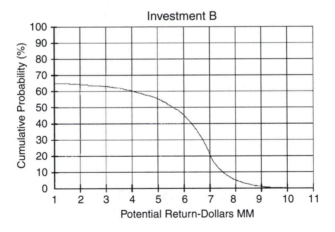

FIGURE 15.2 Cumulative frequency plot for Investment B.

not fail, it will return a minimum of $1 million, a most likely of about $6.8 million, and a maximum of $10 million. Cumulative frequency plots for opportunities A and B appear in Figure 15.1 and Figure 15.2.

Inspection of the plots in Figure 15.1 and Figure 15.2 shows that investment A has a 15% greater chance of succeeding than does investment B (20 vs. 35% chance of failure). If it does not fail, investment A has a higher minimum yield than does B. Likewise, A has a higher maximum yield. On the face of it, then, it might seem that A is the better opportunity. But wait, there is more.

From the cumulative frequency plot of investment B we can see that we have about a 45% chance of getting a return of around $6 million. The probability of realizing a $6 million return from A is less than half that of investment B. Hmm, maybe we should take a look at this in a different space.

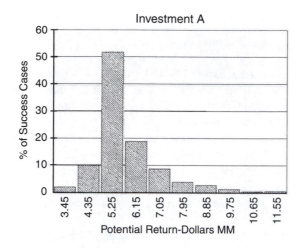

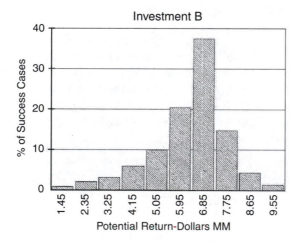

FIGURE 15.3 Frequency plots for two investment opportunities.

In Figure 15.3 are the frequency plots for the two opportunities. Inspection of the plots indicates that the most likely return from investment A, if it does not fail, is about $5.25 million. The most likely return from investment B, if it does not fail, is about $6.8 million. Well, now which investment appears to be the better?

Risk-weighted values facilitate the comparison and ranking of probabilistically assessed opportunities. However, before we use risk-weighted values to select the best opportunity, let's take a look at just how risk-weighted values are calculated. The answer to this problem is given in the Risk-Weighted Values — The Answer section of this chapter.

RISK-WEIGHTED VALUES — HOW?

The fundamental premise behind the calculation of risk-weighted values is simple. In previous sections of this book we have described just how cumulative frequency plots are constructed (see Chapter 11). In Figure 15.4 is shown a set of values that comprise a cumulative frequency curve. Also shown in Figure 15.4 is a blow up of the section of the cumulative frequency curve containing the two points whose coefficients are of greatest magnitude.

The X-axis difference between the two points is 2 (99 – 97). The probability of random selection of the point whose coefficient is of greatest magnitude is one chance out of however many discrete X-axis values there are to choose from (the number of X-axis values is equal to the number of Monte Carlo iterations). Therefore, in the construction of the cumulative frequency plot, the last point is elevated one probability distance above the X axis. The Y axis is subdivided into equal-sized pieces. The number of pieces equals the number of Monte Carlo iterations. One piece is equal to one probability distance along the Y axis. The data value whose coefficient is second in magnitude is elevated two probability distances above the X axis (in a cumulative frequency plot, there are two chances out of the total number of discrete values of selecting the point of second-greatest magnitude or the point of greatest magnitude).

So, we know that the vertical legs of our trapezoid are probabilities and that the base of the trapezoid is the X-axis difference between the two points. We can now calculate the area of this trapezoid and all other trapezoids to the left of this one. The sum of all the individual trapezoidal areas is the risk-weighted value. (We will

Risk-weighted values - how?

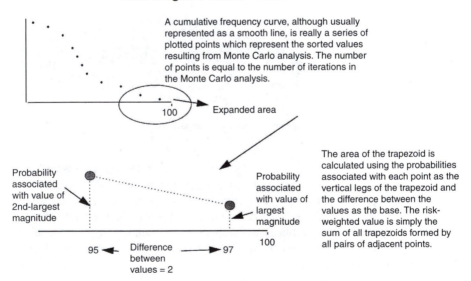

FIGURE 15.4 Plot showing how a risk-weighted value is calculated from a cumulative frequency curve.

see later that if the cumulative frequency plot reaches 100 on the Y axis, then the risk-weighted value so calculated will equal the mean of the X-axis values.)

This methodology, in effect, calculates the area under the curve. Calculation of this area, however, is generally unsatisfactory for many reasons, not the least of which is the fact that the area remains constant regardless of where the curve is situated on the X axis. That is, the risk-weighted value for a curve with a minimum of, say, 10 and a maximum of 100 is exactly the same as the same curve slid down the X axis so that its minimum is 110 and its maximum is 200. The magnitude of the risk-weighted value does not reflect the position of the curve on the X axis. Therefore, some customized risk-weighted values generally are calculated. We will consider two of these.

THE NET RISK-WEIGHTED VALUE

No matter whether a risk assessment begins as a technical, political, environmental, or other type of evaluation, typically the result of such assessments is a monetary or financial calculation. A most useful parameter for comparing and ranking financial risk model output is the net risk-weighted value.

In the previous section it was explained how, in principle, a risk-weighted value is calculated and that calculation of only the area under the curve is not an intrinsically utilitarian measure. If risk-weighted values are to be used to compare and rank multiple risk assessed opportunities, then a mechanism must be employed whereby the changing position of the cumulative frequency curve, relative to the X axis, invokes a change in the risk-weighted value.

For the net risk-weighted value and the economic risk-weighted resource value (to be discussed in a subsequent section of this chapter), we introduce the concept of the limit line. In Figure 15.5 a limit line has been selected that is coincident with

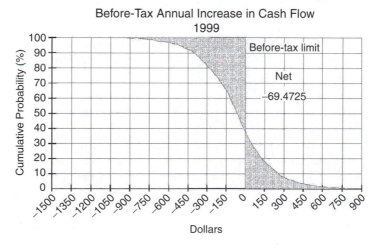

FIGURE 15.5 Cumulative frequency plot. Shaded areas we used to calculated net risk-weighted value.

the Y axis and the X-axis value of 0. The cumulative frequency curve in this figure is typical of many such curves that result from financial risk assessments in that the curve straddles 0. That is, the curve represents both positive (making money) and negative (losing money) values. X-axis values that plot far from 0 in either the positive or negative direction are of relatively great magnitude. The probabilities associated with these values, however, are relatively small. Therefore, on the negative side of the cumulative frequency curve, like their positive counterparts, data points that plot increasingly far from 0 represent increasingly large negative values which also are increasingly less likely to happen in the real world (increasingly small associated probabilities).

As depicted in Figure 15.5, the positive area under the curve is calculated. This is done in the manner previously described. The negative area is calculated in exactly the same way (on the negative side, our trapezoids would be upside down with respect to their positive-curve counterparts). The negative area is then subtracted from the positive area to yield a net risk-weighted area. This risk-weighted measure is a favorite tool of economists and others who must evaluate, compare, and rank financial opportunities. An example of its use will be given at the end of this chapter.

A net risk-weighted value, like any other risk-weighted value, is not an indication of how much an opportunity is expected to produce. That is, the risk-weighted value is not equal to (necessarily) the modal or most likely value on the answer-variable plot. Rather, a stand-alone, risk-weighted value indicates the amount of dollars (or whatever the X axis represents) that the risked opportunity would contribute to a portfolio of opportunities on a fully risk-weighted basis. The best use of a risk-weighted value, however, is as a means of comparison of one opportunity to the next. They are most commonly used as a relative measure. See the section Risk-Weighted Values — The Answer in this chapter for an example.

THE ECONOMIC RISK-WEIGHTED RESOURCE (ERWR) VALUE

When the answer variable is expressed in other than monetary units, such as gallons or barrels of liquid, tons, or other units that generally do not go negative, the economic risk-weighted resource (ERWR) value may be employed. Like the net risk-weighted value, calculation of the ERWR utilizes a limit value — most commonly an economic limit.

Consider the situation in which we are in the midst of a labor strike. We have only enough nonunion workers to operate one of our four factories. In Figure 15.6 is depicted a cumulative frequency curve that represents next month's projected production, in tons, from one of the factories. Values on the X axis indicate that next month's total production from the factory being considered for operation could range from 0 tons (if we, in fact, cannot make it work without union labor help) to 1000 tons. Our economists indicate that if total production is not at least 450 tons next month, then we will lose money for the month.

In response to our economist's calculated lower acceptable limit of production, we have drawn a limit line on the plot at 450 tons. The question we would like

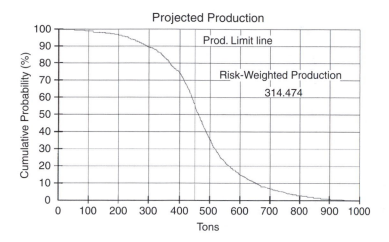

FIGURE 15.6 Cumulative frequency plot of projected production with limit line drawn at 450.

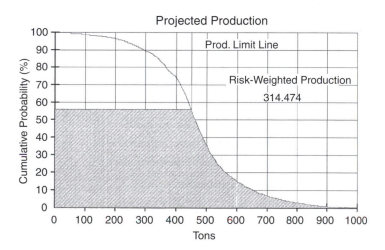

FIGURE 15.7 Cumulative frequency plot of projected production. Shaded area is area used to calculate ERWR (risk-weighted production) value.

answered is, "What is the probability of achieving our lower limit of production, and what is the ERWR value (to be used in comparing one factory's projected output to that of the three others so we can decide which factory to operate)?"

An ERWR value is determined by calculating the area under the cumulative frequency curve up to the limit line and adding to this area the area of the box to the left of the limit line. The box extends all the way to the X-axis value of 0 whether or not it is shown on the plot. The shaded area in Figure 15.7 indicates the ERWR area. This, at first glance (or even at second glance), may seem to be a queer area to serve as the basis for a risk-weighted value.

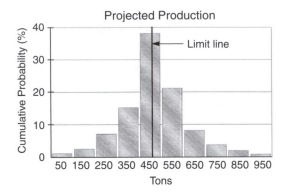

FIGURE 15.8 Frequency plot equivalent to the cumulative frequency plot in Figure 15.7.

Figure 15.8 is the frequency plot for the same data. The area represented by the bars to the right of the limit line in Figure 15.8 is the same area as that shaded in Figure 15.7. The area to the left of the limit line in Figure 15.8 is equivalent to the unshaded area in Figure 15.7 under the cumulative frequency curve. As with the net risk-weighted value, the ERWR is not the most likely amount the factory will produce next month (most likely is somewhere in the vicinity of the tallest bar in Figure 15.8). Rather, it is the amount of production that this factory would contribute to a portfolio of outputs of factories on a fully risk-weighted basis. In the section immediately following, an example is given that makes clear the use of risk-weighted values as ranking tools.

THE EXPECTED VALUE OF SUCCESS (EVS) AND THE EXPECTED VALUE FOR THE PORTFOLIO (EVP)

In previous and subsequent chapters of this book, the concepts of the expected value of success (EVS) and the expected value for the protfolio (EVP) are described and examples delineated. However, given that this chapter is devoted to risk-weighed values, and the EVS and EVP are such values, it would be remiss to fail to at least address the concept here, even though this repeats some descriptions given in earlier chapters. The EVS and EVP concepts had to be related in the earliest chapters because these values form part of the discussions that precede this chapter.

The concepts are simple. The EVS is a coefficient that represents the value of an opportunity. This value can be expressed in any units — tons, barrels, units sold — but is most often expressed in monetary units such as dollars. The EVS resulting from a probabilistic assessment can be the mean value of a range of values. A range of values results from having incorporated uncertainty in the assessment process (i.e., a typical Monte Carlo process where integration of distributions — representing uncertainty for multiple parameters — results in a range of outcomes that typically is portrayed as a cumulative frequency curve). The EVS is the value that results from such an analysis sans the impact of any chance of abject failure. Note that the

"S" in EVS indicates that we have been successful in executing the project and does *not* mean that we have executed a successful project — there is a difference.

A chance of abject failure is the probability that an identified problem will cause benefits resulting from the considered project to fail to contribute to a portfolio of projects in the timeframe considered. For example, the project being considered might be the construction of a chemical plant. We expect the plant to come on line 2 years from now. We expect X tons of production to result in year three. Considering all of the uncertainties associated with plant production in year three, the EVS value is calculated. However, there might be problems that prevent the plant from contributing X tons of product to the corporate portfolio in year three. One such thing might be a late start time for the plant. Corporate officials might know that the competition is planning a similar plant to capture the market. If they break ground first, there is little point in continuing with our plans. A delay of one year or more in beginning our construction would be considered failure because the competition would beat us to the market. The probability that this situation will occur is one chance of abject failure. There could be more such failure modes. Integration of all failure modes is applied to the EVS value to arrive at the EVP. The EVP is the value — this time expressed in tons — that corporate executives would use to roll into the corporate-portfolio projections (for year three) and to compare this project to other projects for the sake of prioritization.

I always allude to the admittedly oversimplified example of the portfolio of 10 dimes lying on a desk. Each dime represents a project that might be executed. Execution of the project is the flipping of the dime. Heads will yield 11 cents and tails 9 cents — each with a 50% probability of happening. The EVS, in this case (5.5 + 4.5) is 10 cents. Independent of this, there is a 50% chance that the project will not be executed at all or will be executed too late. The project then, except for the cost off doing business costs nothing (0.5×0). Each project (dime) has a business unit leader (BUL) and an engineer. The engineer is responsible for building the vessel that will hold and transport the winnings (10 pennies). When the engineer asks the BUL how big to build the vessel, the answer must be that the vessel needs to be big enough to hold 10 pennies. We have to plan for success. If the business is lucky enough to realize success, it will have to be ready to deal with 10 pennies. Logistical contracts and other considerations also are based on the EVS.

However, when the corporate representative asks the BUL how many pennies his project, yet to be executed (i.e., the dime has yet to be flipped), will contribute to the corporate portfolio, the BUL would certainly not reply, 10 pennies. If each of the BULs indicated that the corporation should count on 10 pennies from their project, the corporate projection for the portfolio would be 100 pennies. This, of course, is not correct.

Each project might have a 50% chance of abject failure. Therefore, the risk-weighted value for each project is the EVS (10 pennies) multiplied by the chance of abject failure (50% or 0.5). This results in an EVP for each project of 5 pennies. If BULs report this value as their contribution to the corporate portfolio, the corporation will roll-up 10 5-penny values and will forecast a portfolio value of 50 pennies. This, on the average, is the right value — the EVP for the portfolio.

Of course, in real life, each project will exhibit a unique set of EVSs, unique sets of chances of abject failure, and unique EVP projections. A roll-up of these individual project-level EVP values will result in the risk-weighted corporate projection of portfolio value. Individual project EVPs also can be used to compare projects for the purpose of ranking and prioritization. An example of application of these principles is given in the final chapter of this book.

RISK-WEIGHTED VALUES — THE ANSWER

As indicated in the immediately preceding sections of this chapter, individual risk-weighted values do have meaning in the context of portfolio analysis. The real power of risk-weighted values, however, is mainly realized when they are used to compare opportunities.

Let's return to the original problem presented in the Risk-Weighted Values — Why? section of this chapter. This problem presents us with the task of choosing between two investment opportunities. Probabilistic analysis of the two opportunities yields the plots shown in Figure 15.1 and Figure 15.2. The X axes of the cumulative frequency plots for both opportunities cover different ranges. The curve shapes are different. The Y-axis intercepts (chance of failure) of the two investments are not equal. Other parameters such as economic limits could be different (but are not in this example). How does a person integrate in their minds changing ranges, changing curves, changing intercepts, etc.? Such integration is exactly the service provided by risk-weighted values.

Because investments generally are considered relative to 0, the limit lines used in the calculation of the net risk-weighted values in Figure 15.9 and Figure 15.10 are at 0. The net risk-weighted value for investment A is 4.597. That for investment B is 4.125. Because the X axes of these plots represent dollars to be gained, then investment A is the best choice (all other things being equal, of course). If the X axis represented dollars to be paid, such as costs, then the smaller value would be the best choice.

SENSITIVITY ANALYSIS — WHY?

The purpose of most risk analyses is to ascertain the validity and quality of one or more opportunities. Most comprehensive risk studies are composed of many, sometimes dozens, of input and output variables. These variables are linked by a system of equations that typically integrate technical, political, financial/commercial, and other parameters.

Each opportunity is assessed by the risk model with the expectation that output from the model will reveal the opportunity to be of good or poor quality — that is, an opportunity you might pursue or one you should forego. If risk-model results indicate that a particular opportunity is of high quality, you should endeavor to determine which risk-model input parameters contribute most to its relatively stellar standing. For example, it may turn out that an investment opportunity ranks high primarily because of the quality of the labor force (as opposed to favorable interest

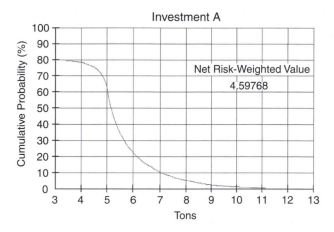

FIGURE 15.9 Cumulative frequency plot for investment A and net risk-weighted value.

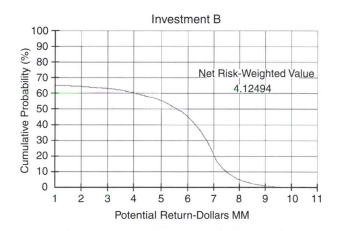

FIGURE 15.10 Cumulative frequency plot for investment B and net risk-weighted value.

rates, quality raw materials, or other risk-model input parameters). Steps should then be taken to preserve the quality of the labor force so that the quality of the opportunity can be maintained. This is not to say that other parameters also are not important; however, there is generally only so much money, time, and effort that can be directed toward any given opportunity. Therefore, it is essential that we discover and nurture the most important contributing variables.

Conversely, the assessment of an opportunity may reveal it to be of poor quality. However, for political or other reasons we may wish to pursue the opportunity if its quality can be improved. A given opportunity may contain political, financial, environmental, commercial, technical, and other input variables. These parameters typically serve as the terms in complex equations that integrate them. If the quality of an opportunity is to be enhanced, it is essential that we discover which parameters, if improved, would result in the greatest increase in quality of the opportunity. With limited resources it is beyond our ability to fix everything, so it is essential that we attempt to fix the parameters that will most improve the overall quality of the opportunity.

Inspection of the system of equations that comprise a risk model usually is of little or no help when attempting to identify critical variables. Unless the equations are simple sums or other equally credible notations, perusal of the lines of computer code ordinarily will not reveal the magnitude of influence associated with individual variables. Neither can such influence, again with the exception of the most simple equations, be deduced from the magnitude of the coefficients which represent the variables. For example, if in a Monte Carlo-based risk model we consider the simple equation

$$C = A + B \qquad (15.1)$$

in which A is a distribution with a minimum of 10,000,000 and a maximum of 100,000,000 while B is represented by coefficients that range from 2 to 6, it should be obvious that the value of C is overwhelmingly influenced by variable A. Variable B exhibits little influence. Variable A, therefore, is the controlling parameter. However, if we change the equation to

$$C = A^B \qquad (15.2)$$

then it is clear that the influence of B on the value of C has dramatically increased even though the magnitude of the coefficients representing A and B are unchanged. Therefore, except in the most mundane of circumstances, it is folly to suppose that the variables whose coefficients are of greatest magnitude also are those which exert the greatest influence on the answer variables.

SENSITIVITY ANALYSIS — HOW?

Through the years, many attempts have been made to develop algorithms which will, for a complex set of equations, reveal which of many input variables are those of greatest influence on the output variables. One of the more popular but significantly flawed methodologies is to hold all variables, save one, at a constant value in order to determine what influence the variable not held constant has on the coefficients of the output variables. It should not take a rocket scientist, however, to realize that the true influence on an output variable depends on the interaction of all coefficient ranges for all variables and that the hold all but one constant (HABOC) method might not, in most cases, reveal the true influence exerted by any variable. Given the popularity of this approach, it is worth a few paragraphs here to describe some of the drawbacks to the hold all but one constant method.

First, I would like to state that there is nothing intrinsically amiss with the form of a tornado diagram. If, for example, we have measured or calculated the ranges of net incomes of various businesses, there is not a thing wrong with displaying those net income ranges as a tornado plot. The range in net income for each business can be represented by a bar of appropriate width and placed on the diagram accordingly. Such tornado plots are informative and aesthetically satisfying. So, the question concerning tornado diagrams is not whether they are truthful or useful, but rather how the information displayed was calculated.

In a risk model, the coefficients for all variables change with each iteration. It is difficult, by inspection, to ascertain the contribution of a single input variable to the change (relative to the previous iteration) in the value of an output variable. To remedy this, the HABOC methodology was born.

Using this approach, in order to determine the influence of a single input variable on an output variable, all variables, save one, are held at a constant value. The value used as the constant typically is a value determined to be representative of the distribution. Mean values, median values, and those termed deterministic are commonly employed. The Monte Carlo process is imposed, holding all variables constant save one that is allowed to vary randomly throughout its range. Each variable has its turn to be allowed to vary while all others are held constant. For each variable allowed to change, the range of the output variable is noted, and this becomes the width of the bar for that input variable in the tornado plot. For speed (or for lack of sophistication), some approaches use only, for example, the 10th, 50th, and 90th percentile values for the variable that is allowed to change value. Discretization only worsens the problem.

The first thing that should strike you as suspicious concerning this methodology is that it violates dependence (see Chapter 14) and creates illegal combinations. For example, consider the two variables, depth and temperature, in a risk model. In the real geological world, these two variables are related. As depth (in the earth) increases, so does temperature (the deeper you go, the hotter it gets). This relationship is termed the geothermal gradient and varies with geographic location.

The depth distribution might contain values ranging from 0 feet to 20,000 feet. Concomitant temperatures might range from 50°F to 500°F. Using the HABOC method, if you are attempting to determine the influence of depth on an answer variable, you would allow depth in the Monte Carlo process to vary throughout its 0 to 20,000 range while holding temperature constant at its deterministic value. Because I have not here displayed either the depth or the temperature distributions, allow me to suggest that the deterministic value for temperature is 220°F. (It does not matter what number you choose as the constant temperature value — the result is the same.)

Using the HABOC technique, depth is allowed to vary while temperature is held constant. So, depth values of 100 feet are combined in the equations with the constant 220°F temperature as are depth values of 19,000 feet. These and countless other depth/temperature pairs are, as far as nature is concerned, illegal combinations. Yet, the variation of the answer variable is still, under this scheme, considered to reflect the influence of depth on that variable. I do not think so.

Still another and more insidious drawback to the HABOC method is linked to branching in the risk-model logic. Consider the situation in which we have two variables A and B that are combined to create an output variable C. Variable A ranges, say, from 0 to 100. Consider also that our equations for calculating C contain a probabilistic branch thus:

$$\text{If } A < 50 \text{ then}$$

$$C = B + 1$$

Else

$$C = B^{100} \tag{15.3}$$

When attempting to determine the influence of B on the answer, the HABOC logic would have us hold A constant at some deterministic value. Again, it is irrelevant what value we select as the constant for variable A — let's pick 55. So, on each iteration the model will execute the if/else logic and always calculate C to be equal to B^{100}. The C value of B + 1 will never be executed. The bar width indicating the influence of B on C will be drastically exaggerated because if A were allowed to vary, the equation B + 1 would be executed a statistically significant number of times. Therefore, the bar representing the influence of B on C is grossly in error. There is a better way.

To date, the most satisfying methodology developed for the purpose of discovering the relative influence of input variables upon the magnitude of coefficients for output variables is the Spearman rank correlation (SRC) method. In keeping with the practical bent of this book, a detailed description of the SRC equations was deemed unnecessary dunnage. Complete disclosure of the SRC methodology can be found in nearly any textbook on basic statistics (see the selected readings section for this chapter). However, it would be careless to refrain from putting forth at least a cursory description of Spearman's correlation method.

SRC is a nonparametric (distribution-free) technique that facilitates the calculation of correlation while allowing all parameters in a multivariate analysis to vary simultaneously. Correspondence between individual input variables and a calculated output variable is arrived at by comparison of ranks. Ranks are employed in situations in which actual measurements are impractical or impossible to obtain. A mundane example is an ice-skating contest in which judges assign subjective scores. The scores are used to rank the skaters. In a two-judge (1 and 2) and three-skater (A, B, and C) contest, judges may both rank the skaters in the order B, C, A. In this case there is perfect correspondence (+1) between the rankings. If judges 1 and 2 had ranked the skaters in exactly opposite order, the rank correlation would be −1.

If measurements rather than ranks comprise the original data set, then the measurements must be converted to ranks before SRC can be applied to the data. Conversion from measurements to ranks likely will result in the loss of some information. Therefore, it is unrealistic to expect that the product–moment correlation coefficient and that resulting from SRC will be equal when both methods are applied to a common set of data.

Loss of information notwithstanding, SRC may be, relative to product–moment correlation, a superior measure of correspondence. SRC's superior position is derived from the fact that it is not predicated on the usually unwarranted presumption that the coefficients of the input variables are normally distributed. Correlation between variables representing coefficient distributions of nearly any type can receive valid treatment by the SRC technique.

SENSITIVITY ANALYSIS — RESULTS

Output from SRC analysis is commonly displayed as a tornado diagram such as that shown in Figure 15.11. On a scale of −1 to +1, the length of the bars indicates the degree of correspondence between a given input variable and the calculated output variable. A separate tornado plot must be produced for each output parameter.

In the plot shown in Figure 15.11, the numbers in parentheses adjacent to the variable names indicate the percentage of variance of the output variable that is explained or accounted for by that input variable. Bars extending to the right of the 0 center line are variables that are positively correlated with the output variable (as coefficients for the input variable increase in magnitude, so do the coefficients of the output variable). Bars extending to the left of the center line indicate negative correlations.

Sensitivity plots like the one shown in Figure 15.11 are not only fundamental to determining which are the prominent input variables, but can be invaluable indicators of whether a particular project should be pursued. For example, the results from a risk model may indicate that we likely will lose money if we pursue the risk assessed opportunity. For political reasons, however, we would like to go ahead with the project if the losses are not significant. We hope that our sensitivity analysis will reveal which of the aspects of the project we might seek to improve, thereby enhancing the palatability of the endeavor.

Our plot may indicate, however, that the two most significant variables, accounting for the majority of the variance of the output variable, are parameters over which

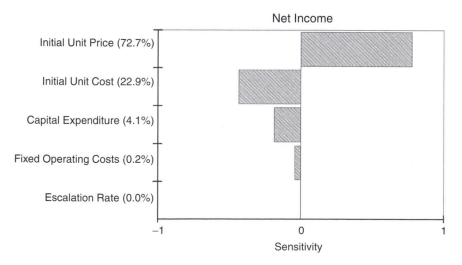

FIGURE 15.11 Sensitivity plot.

we have little or no control. Parameters such as market price and taxes are good examples. As in this case, if parameters beyond our control are the significant factors in a discouraging risk assessment, then the sensitivity plot is not only an indicator of which variables are significant, but is also an indicator of whether it is logical and beneficial to pursue an opportunity.

SELECTED READINGS

Ebdon, D., *Statistics in Geography — A Practical Approach*, Basil Blackwell, Oxford, UK, 1981.

Saltelli, A., Chan, K., and Scott, E. M., *Sensitivity Analysis*, John Wiley & Sons, Hoboken, NJ, 2000.

Saltelli, A., Tarantola, S., Campolongo, F., and Ratto, M., *Sensitivity Analysis in Practice*, John Wiley & Sons, Hoboken, NJ, 2004.

Welch, David A., *Decisions, Decisions: The Art of Effective Decision Making*, Prometheus Books, Amherst, NY, 2001.

16 What Companies Really Want to Do

THE PORTFOLIO EFFECT — SOMETIMES YOU JUST DO NOT GET IT

It is good advice to diversify. Originally espoused formally for investment portfolios by Harry Markowitz, it is advice intended to allow you to reduce your portfolio's exposure to risk. Boiled down to its basic premise, this philosophy simply proposes that if your portfolio is comprised of a diverse set of investments, it is unlikely that all investment types will experience a downturn (i.e., a loss of value) simultaneously.

This type of investment advice is fine for the typical investor who is working with a portfolio of opportunities (managing the distribution of funds in your 401K retirement plan, for example). However, many of our most vexing corporate investment decisions are those that do not directly involve portfolios. Should we continue to fund this mature project? How much money should be earmarked for stock buyback? Should we invest heavily in a foreign joint venture? Questions such as these are ubiquitous and practical — most corporations wrestle with such quandaries related to the value of investments.

Even vaunted applications such as the Black-Scholes model are of little help in resolving such problems. It is true that the Black-Scholes equation is a great application relative to previous attempts to solve the option-pricing problem. The equation is a triumph of simplification, lending credence to the proposition that better does not necessarily relate to more complex. It does, in fact, address the problem of how to determine the value of a single option (an investment to be taken in the future). However, the calculated price determined for an option could turn out to be anomalously high or low. Protection against such miscalculations is mainly provided by making such calculations on a portfolio of options. The diversification principle, hopefully, should in the portfolio ensure that overestimations damp the impact of underestimations and that the desired return is realized.

Although such instruments as the Black-Scholes model are relatively simple and useful, such equations typically utilize parameter coefficients that de-emphasize the insight, goals, feelings, and fears of corporate executives for whom the calculations are made. For example, the Black-Scholes model includes parameters such as the call option price, the current stock price, the exercise price, logarithms, time remaining to the expiration date, cumulative normal probability, and a few other hard variables. However, most individual investment decisions involve such corporate-specific questions as the following:

- Just how many more years do we expect this project to be profitable?
- Given the product produced by one arm of the company and our opinion regarding the competition, just how much longer do we think the company's market share can be maintained?

Answers for these questions and thousands more like them require integration of hard data such as that demanded by the class of calculations exemplified by the Black-Scholes model, and soft data such as a corporation's opinions, requirements, goals, obligations, fears, and other corporate-specific information. Melding these diverse information types with customized stochastic (iterative and probabilistic) models is the crux of successful execution of the risk/uncertainty process. The portfolio effect related to diversity does not much help in making decisions on individual investments.

Procedures outlined in this book address the problem of integration of hard and soft data in a comprehensive model. For example, the epitome of a soft problem is that of integrating political considerations into, for example, a projected financial benefit related to a project. Hard data such as capital expenditures (CAPEX), operational expenditures (OPEX), and other data can be gathered. Consideration of a political problem — such as a government's delay in granting essential permits — might need to be integrated. Just one way to approach this is to consider the following:

- The probability — expressed as a percentage — of the political problem happening
- The duration — expressed in time units (days, weeks, etc.) — of the problem
- The cost per time unit — expressed in monetary units (dollars, for example)

So, if we were uncertain about any of the parameters above, we could generate the appropriate distributions (for example, one in percent, one in days, and one in dollars) and simply multiply them together to get the impact of the political problem.

Political Problem Impact = (% probability) x (days delay) x (dollars per day)

$$(16.1)$$

This impact can then be easily integrated with our hard data to arrive at a total financial benefit for the project.

Successfully performing such an integration for each project in a portfolio, however, does not solve the problem of achieving the best mix of projects in the optimal portfolio. Too often, corporations are siloed. That is, individual business units work independently and put forward their best projects. It would be an accident indeed if the collection of the best projects from each business unit comprised the optimal portfolio.

It is beyond the ken of this book to suggest how any corporation should perform its portfolio-analysis process. However, it should be recognized that simply because comprehensive risk processes might have been utilized on each project in each disparate business unit, the collection of the best projects from each unit very likely does not make the best mix of investments. It might, for example, be smarter to fund no projects from business unit X and to utilize the funds that might have been used to execute business unit X's best project (or projects) to fund multiple projects from business unit Y.

I have never known any corporation to use algorithms to achieve the best mix of projects in a portfolio. It would seem logical that if all projects had been subject to a comprehensive risk analysis and a common risk-weighted measure (NPV, for example) had been generated for each potential investment, it would be a simple thing to design an algorithm that would, in the end, provide the optimal mix of investments. I have seen such algorithms. I have never witnessed their adoption as the ultimate decision-making process with regard to selecting projects to comprise a corporate portfolio.

Why not? Several reasons, really, but the most salient is that corporate executives very much want to impose their individual and collective wisdom and experience on the process of project selection. Timing issues, political concerns, personality considerations, and other items that are poorly dealt with in an algorithmic sense come into play. It is not so much that an algorithm cannot be built to address, for example, political, timing, and personality considerations, but it is supremely difficult to encode the integration of such considerations that is practiced in one person's mind, let alone the integration process that results from group dynamics associated with a cadre of decision makers.

So, what is the answer? All I can do is relate to the reader my experience. I have found that it is best to have performed comprehensive and easily-explained risk analyses on each project to be considered. Such analyses should have resulted in at least one common risk-weighted metric so that projects can be compared and ranked. These metrics and other information should be presented in a common and easy-to-digest format. Then, it is up to the decision makers to consider how they will integrate the data and rank the projects to arrive at what they consider to be the best mix of investments for the portfolio. Most attempts to supplant experience and wisdom with mathematical rigor are bound to be given only cursory consideration.

One last comment regarding creation of a best portfolio. Most valuation methods are designed to minimize or remove the consideration of time from the potential investment. Bringing time specifically into the analysis process is not supremely difficult. Discounted cash-flow analysis is just one example of how time-consideration of money can be achieved in the probabilistic risk analysis process. However, presentation of the time dimension in a simple way in an executive summary of project assessments is, in fact, a very daunting problem.

I have found it best to address this problem in two interrelated ways. First, select at least one metric of project value that specifically addresses the time element. For example, net present value (NPV) can be the result of a time series of probabilistic discounted cash flows. This might not address the issue of relative timing of projects (which project should come first, which one second, etc.), but it does at least bring the time element into the conversation. NPV is not the sole metric that specifically addresses timing. There are many metrics from which to choose, and each company should select the measure that best fits its needs.

A second way to address the timing issue is to be sure that it is simply part of the conversations that take place when the decision makers are using their experience and wisdom to select a portfolio of projects. Part of the simple-to-digest information related to the project should be an expression of the time it will take to execute the project and any other issues related to time. It might seem like I am punting here and passing the problem along to the decision makers — and in effect, I am. However, it has been my experience that there is no better means of dealing with the complexities of project timing in a portfolio than to foster the appropriate conversations among the decision makers. If there is a better and more practical means to address timing in a portfolio (and do not tell me about algorithms), I do not know what it is.

ESTIMATE BUDGETS

Both long- and short-term budget estimating is at least an annual event in most corporations. Typically these budgets are spreadsheet-based roll-ups of a series of deterministic estimates. Also typical is sandbagging and other manipulations of the numbers submitted. Although there is no sure cure for people gaming the system, a comprehensive probabilistic budget-estimating process can resolve many problems. In addition, the probabilistic budgetary process explained in this section will be shown to generate budgetary values that much better match actual annual and long-term spending. Other problems related to corporate attitudes toward underspending, and the ramifications of setting estimates that are relatively accurate, will be addressed. This is a complex issue; however, a practical and easily implemented process will be detailed in this section.

In the past, I considered myself to be ever so smart and accomplished when I could persuade a business entity to abandon deterministic assessment processes for those that considered uncertainty and ranges — especially when this related to budget estimates. Come to find out, embracing the uncertainty-related approach did not really change final estimates much relative to the deterministic method. "How can this be?" you say.

Consider the portfolio of 10 projects shown in Figure 16.1. Each value listed is the expected cost for the project for the next budget year. The sum of the projected project costs is the amount of money for the year that would be requested to execute this portfolio of projects.

Now, consider that we were successful in convincing project leaders that they are really uncertain about the costs for the projects and that ranges should be considered. The values shown in Figure 16.2 represent the ranges in cost for each project following our in-depth conversations. Using a Monte Carlo model, we can combine the ranges to get a total annual budget range for the portfolio. This plot is shown in Figure 16.3. The mean value is shown on the plot.

Even though the plot in Figure 16.3 does represent the range of probable costs for the project portfolio — and this is useful information — the thing that people seem to focus most on is that the mean value is very similar to the sum of the most

Simple Portfolio Example

Project	
Project 1	3
Project 2	9.2
Project 3	3.5
Project 4	10.3
Project 5	4.6
Project 6	6.8
Project 7	5.4
Project 8	15.1
Project 9	5.5
Project 10	12.4
TOTAL (SUM $MM)	**75.8**

FIGURE 16.1 Deterministic costs for ten projects.

Simple portfolio example

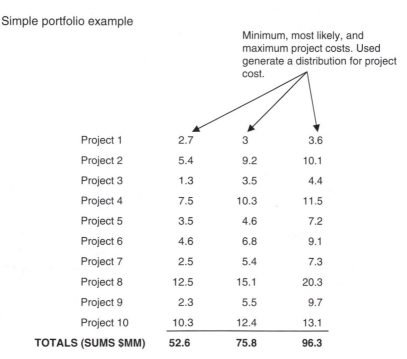

Minimum, most likely, and maximum project costs. Used generate a distribution for project cost.

Project 1	2.7	3	3.6
Project 2	5.4	9.2	10.1
Project 3	1.3	3.5	4.4
Project 4	7.5	10.3	11.5
Project 5	3.5	4.6	7.2
Project 6	4.6	6.8	9.1
Project 7	2.5	5.4	7.3
Project 8	12.5	15.1	20.3
Project 9	2.3	5.5	9.7
Project 10	10.3	12.4	13.1
TOTALS (SUMS $MM)	**52.6**	**75.8**	**96.3**

FIGURE 16.2 Ranges of costs for ten projects.

likely values. Most budget processes require a single value to be reported, whether or not it was the result of summing most likely values or whether it was a value selected from a range of values such as the mean shown in Figure 16.3. Therefore, the comment almost invariably made is this: "It is nice to know the range, but given that the mean value from the range is usually similar to the sum of our most likely values, why go through all the trouble to perform Monte Carlo analysis if we are only going to pick the mean off the curve — which is essentially a value we could have without resorting to the Monte Carlo process?"

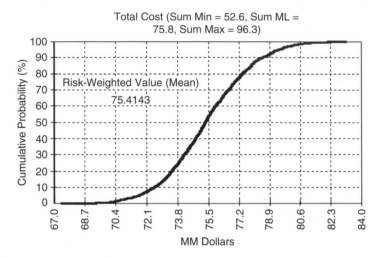

FIGURE 16.3 Cumulative frequency plot of total budget cost considering only ranges of costs.

I found it of little solace to try to convince folks that the mean value matching the sum of the most likely values is an artifact of distribution shapes, etc. This was wasted effort. However, serendipity came to the rescue.

Most budget estimates for such portfolios tend to be high. That is, the amount of money forecast to be needed to fund the portfolio — if the individual project-cost estimates are honest — is greater that the amount that will be spent in the given year on that portfolio. This is true whether or not the sum is derived by summing most likely estimates or by calculating a mean from a distribution as in Figure 16.3. It should be noted here that the author is aware that only mean values can be legally summed, but I have never known that fact to hinder folks from summing minimum values or maximum values or whatever they consider to be most likely values. So, although we know this practice is illegal, the reality is that it is done all the time.

The budget estimates turn out to be high mainly because not every project in the portfolio will be executed in the budget year. That is, some projects will be dropped from the portfolio or the execution of a project will be delayed — sometimes partially, sometimes completely — until a time outside the range of the current budgetary consideration. Political problems, technical setbacks, environmental considerations, and other project-schedule problems can and do push project spending outside the current budget window.

So, a variation of the chance of failure concept needs to be applied. Note that in Figure 16.4 there is an additional column of information. Values in this leftmost column represent the percent chance that the project *will not* contribute to the portfolio costs in the timeframe considered. Therefore, a 0 (zero) indicates that there is no chance that this project will fail to contribute costs to our sum of costs. A 90 value indicates that 90% of the time this project will fail to contribute to our sum of costs.

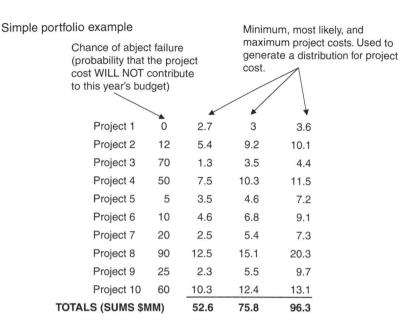

Simple portfolio example

Chance of abject failure (probability that the project cost WILL NOT contribute to this year's budget)

Minimum, most likely, and maximum project costs. Used to generate a distribution for project cost.

Project 1	0	2.7	3	3.6
Project 2	12	5.4	9.2	10.1
Project 3	70	1.3	3.5	4.4
Project 4	50	7.5	10.3	11.5
Project 5	5	3.5	4.6	7.2
Project 6	10	4.6	6.8	9.1
Project 7	20	2.5	5.4	7.3
Project 8	90	12.5	15.1	20.3
Project 9	25	2.3	5.5	9.7
Project 10	60	10.3	12.4	13.1
TOTALS (SUMS $MM)		**52.6**	**75.8**	**96.3**

FIGURE 16.4 Ranges of costs and project-associated chance of abject failure (i.e., chance of project not contributing to the total budget in the timeframe considered).

On each iteration of the Monte Carlo model, the program generates a separate random number between 0 and 100 for each project and tests it against the percent probability for that project. If the randomly-generated value falls between 0 and the percent probability for that project, then that project does not contribute a cost to sum representing the portfolio cost. If the randomly-selected value between 0 and 100 is greater than the percent probability for that project, then a cost — drawn at random from the distribution of costs for that project — is used in the sum representing the total portfolio cost.

A plot of total costs for the portfolio — using the chance of failure process — is shown in Figure 16.5. Note that the mean value is less than the sum of the minimum values. (Yes, I know, you should not sum the minimum values.) This usually gets people's attention. Note that the maximum value of the X-axis range for this plot is near in magnitude to the mean of the values resulting from combining individual project-cost ranges when the chance of failure process is not used. This indicates that in the past, there was little hope that spending for the portfolio would actually come close to a budget projection sans the chance of failure process. Comparison of actual annual spending with chance-of-failure-enhanced budgeting has been very favorable.

Note two things. The process used in this simple example utilized deterministic estimates of probability for each project. I could have just as easily used ranges of percentages to represent the uncertainty related to whether or not a project would contribute a cost to the sum of portfolio costs in the timeframe considered. For the sake of simplicity of demonstration, I chose to use single-valued percentages. Also, this process is universally applicable. That is, there is not a part of a

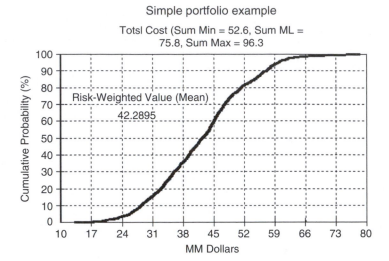

FIGURE 16.5 Cumulative frequency plot of total budget cost considering ranges of costs and chance of abject failure.

corporation that can not benefit from incorporation of this method — or a twist on this method — in their budgeting process.

ESTIMATE PROJECT VALUE — TWO TYPES: THE EXPECTED VALUE OF SUCCESS (EVS) AND THE EXPECTED VALUE FOR THE PORTFOLIO (EVP)

I have addressed the expected value of success (EVS) and expected value for the portfolio (EVP) concepts in earlier chapters of this book. However, they bear mentioning here again primarily because they relate directly to what corporations should do. An example of the application of these values is given in the following chapter.

In most corporations, a single number is generated to represent the value of a project (value can be expressed as dollars, tons, barrels, or in other units). Most probabilistic processes, in the end, have the effect of converting that deterministic estimate of project value into a range. This step, however, represents only half the battle.

Two measures of value are required for most projects — the expected value of success (EVS) and the expected value for the portfolio (EVP). The simplest analogy for this concept is the portfolio of 10 dimes lying on a table. Each dime represents a project. Execution of the project is represented by the flipping of the dime. Heads will yield 11 cents and tails 9 cents — each with a 50% probability of happening. The EVS, in this case (5.5 + 4.5) is 10 cents. Independent of this, there is a 50% chance that the project will not be executed at all or will be executed too late. The project then, except for the cost of doing business, costs nothing (0.5×0). The EVP results from combining the EVS with the impact resulting from abject failure. In this simple case, the EVP is 5 cents.

For any given project (dime) our engineer might come to us and ask: "How big a penny roll do I need to build to hold our potential winnings?" We have to answer that the penny roll should be built to hold 10 pennies because if we are lucky enough to experience success, we had better be able to handle it. An energy company analogy would be the 100 MMBOE (millions of barrels of oil equivalent) subsurface prospect. There might be a 50% chance that the prospect will be a dry hole (i.e., no economic hydrocarbons). When the engineer comes to us and asks how big a gathering system and processing facility he should plan to build, we have to answer that he should build one big enough to handle 100 MMBOE, because if we do successfully tap the reservoir, that is what is going to issue forth from the earth and we need to be ready for it. This value is the expected value of success — the EVS. We build for it, we sign logistical contracts based on it, and so on.

So, to return to the portfolio of dimes analogy, the EVS is 10. That is, if we execute the project, we will get 10 pennies, and we had better be ready to handle success. Corporate management, however, wish to report just what levels of income we are expecting from our portfolio. We certainly would not add up all of the EVS values (10 10s) and announce that we expect our corporate portfolio to yield 100 cents. We might know that each project has a 50% chance of abject failure. So, to get the EVP, we modify the EVS by the chance of abject failure for each project (0.5 x 10 = 5) to get an EVP for each project (i.e., each dime) of 5 cents. The portfolio, then, is worth, on the average, 50 cents — the sum of the individual-project EVP values. Of course, in any actual corporate portfolio, each project would likely have a unique chance of failure (not all the same — 50% — as in the dime analogy), but the principle prevails.

Both these values — the EVS and the EVP — are exceedingly important in the corporate world, and each has its use. Generation of these values in real-world corporate environments and the correct use of these values is the challenge. It can be a daunting task to get any business to begin to generate any sort of expected value (EV). It is an especially horrendous task to convince businesses that two such values should be created for each project or potential investment. In the following chapter, a comprehensive example of creating and using the EVS and the EVP is given.

ASK THE RIGHT QUESTIONS TO DETERMINE THE QUALITY OF THE INFORMATION BEING CONVEYED

It is sometimes true that new proposals are shot down because those to whom the proposals are presented simply do not know what questions to ask in order to reasonably evaluate the new offering. This is more common than the reader might imagine.

Most individuals who are decision makers reached the position of decision maker because they are relatively astute. These people also are exceedingly busy and tend to rely on the processes and techniques that served them well in their careers on the way to the decision maker position. For a new proposal to be accepted that purports to change the way he thinks about or evaluates things, the decision maker needs to believe that the new method or technique actually has merit and is worthy of adding to or modifying their tried-and-true experience-based methods.

How do they get that assurance? The three questions proposed and discussed below certainly are not the only queries relevant to this problem. However, these three questions can form the basis for a informed conversation regarding the relative merits of a new proposed method and why it should be adopted as part of the already established decision-making process.

How Does This New Information Impact the Decision or the Decision-Making Process?

As mentioned previously, decision makers are not looking for a more cool way to make calculations or new and trendy metrics. These folks honestly want to make the best decisions possible given the information that can be practically had and digested.

If the statements above are accepted, then it should be clear that when a new process or technique is proposed — such as probability-based risk methods — the emphasis decidedly should *not*, except where necessary, be on the technology behind the new process. Focus should be put on how such techniques and their output can positively and practically impact the established decision-making process.

For example, it is important to emphasize — and this has to be true — that the proposed methodology is not an undue burden on the businesses and project teams that have to generate the probabilistic assessments in the first place. Examples of real application and testimonials from business-unit leaders regarding the positive aspects of the process are essential.

In addition, it could be emphasized that due to the nature of the probability-based process (determining what is the question, creating contributing-factor plots, discussing ranges and dependencies, and other characteristics of the process outlined in previous chapters), decision makers can be more assured of the following:

- The values that they are comparing from project to project are reasonably consistent and comparable.
- Options and ranges have been considered for each project.
- Systemic as well as independent impacts have been considered.
- The values they will use for comparison result from a holistic assessment of the problem including soft risks and uncertainties (i.e., reputation issues, political problems, legal considerations, etc.) as well as hard attributes such as cost, schedule, markets, etc.
- The values they are using for evaluation of a project or comparison of multiple projects are values that are likely as results. They can know where these values stand in the range of possible outcomes and the probability of achieving any given value.

As mentioned previously, it is not the best use of time and effort to argue that the numbers that result from having implemented the risk-based approach are superior to those that would result from a process sans the risk aspect. One cannot know the road not taken and it is futile to argue against the person who claims: "I would have arrived at the same conclusion — cheaper and faster — if I had not gone through all the trouble of implementing this risk stuff." There is really no practical way to disprove this person's argument or to disprove an argument that claims the superiority of single values that might be drawn from a probabilistic process. However, it certainly can be argued that the process leading up to those numbers can and does help both the project teams and the corporation. The best results from probabilistic process implementation are typically the conversations between project team members — especially between seemingly disparate groups such as, for example, lawyers and engineers — and the modification of behaviors of the project team members. When behaviors are changed such that project personnel consider options, ranges, a holistic approach, and the like, better and more meaningful conversations result. In turn, better decisions are likely.

What Assurance Do I Have That the Process That Generated This New Information Is Practical and Credible?

The answer to this question is straightforward. No amount of theoretical discussion is likely to persuade decision makers to wholeheartedly adopt a new method — especially if that new method supplants aspects of a tried-and-true and trusted process.

Assurance stems from actual application and testimonial. So, the proponent of the new risk-based process who would like decision makers to consider probabilistic attributes in the decision-making process has to demonstrate how the risk-based approach was of benefit to the analysis of real business cases. Real results that clearly show how a project team, for example, better analyzed a problem — relative to the old ways — are required.

For example, a project team might contrast the previously utilized deterministic process with the new risk-based process. Under the old method, each entity — marketing, research, legal, engineering, logistics, etc. — was expected to contribute its estimates to the overall project economic model. This method might be contrasted with the risk-based approach in which a holistic and integrated method is prescribed. Under this method, various project-team factions are required to engage in conversations that uncover new problems and opportunities and point out synergies that would previously have gone undiscovered. Great emphasis should be put on the improved dynamics of the project team when the new process is utilized.

Testimonials also are important. Project and business leaders should be enlisted to relate the unvarnished story. Certainly, along with the positive aspects of implementing the risk-based approach, the downsides of the method also should be conveyed, including extra costs and time required, training of personnel, organizational issues that arose, and so on. Credibility is damaged when only the joyous ramifications are presented. If the downside aspects are related and it is indicated that those problems were handled, astute decision makers are better able to accept the new proposal. They know that no change is free, and when they understand and accept the costs, they are better able to embrace the benefits.

How Can This New Information Be Integrated Into the Existing Decision-Making Process?

Upon receiving new risk-related information, it is unrealistic to expect that decision makers are going to supplant their established decision-making methods entirely with a new risk-based process. The new information has to fit into the established decision-making framework and add value.

For example, an established corporate decision-making method might include the comparison of deterministically determined measures of project value — say, single-valued NPVs. Typically, such metrics result from spreadsheet-based economic models. Such models take in single values in cells and report single values as results. To enhance decision making, a new probability- and Monte Carlo-based process might be proposed. Such a process yields as output ranges (i.e., cumulative frequency and frequency curves) probabilities associated with each potential outcome, sensitivity analyses, and the like. However, it will likely not be clear to decision makers just how all this fits into their existing decision process.

The savvy proponent of the risk-based process might approach the problem in this manner:

- Show that a single value (mean or other selected value) can consistently be drawn from the NPV range for use as the single value which the current decision process demands.
- Show that knowing the probability of achieving that value is a very useful piece of information. Using the spreadsheet-based deterministic estimates of NPV, decision makers could not be sure whether such a value was too low a target, too much of a stretch target, or where in the realm of likely outcomes such a value lay — even when multiple deterministic what if scenarios were run.
- Show that there is merit in knowing the range of likely outcomes. This allows decision makers to set reasonable stretch targets and understand the potential downside of selecting a particular project.
- Demonstrate that combining the outcomes from multiple projects is as easy as using a Monte Carlo model to sum the individual output curves from each project. A single value can then be drawn off the sum curve for use in decision making and reporting.
- Sensitivity analysis can indicate where the corporation or individual business should concentrate its efforts and spend to create a more certain result for any given project or portfolio of projects.

When decision makers realize that adopting the probabilistic method *can* fit into their existing process and that many other useful attributes are available, they are much more likely to embrace the new technology and method.

SELECTED READINGS

Campbell, J. M., *Analyzing and Managing Risky Investments, 1st Edition*, John M. Campbell, Norman, OK, 2001.

Chriss, N. A., *Black Scholes and Beyond*, McGraw-Hill, New York, NY, 1996.

Elton, E. J., Gruber, M. J., Brown, S. J., and Goetzmann, W. N., *Modern Portfolio Theory and Investment Analysis, 6th Edition*, John Wiley & Sons, Hoboken, NJ, 2002.

Koller, G. R., *Risk Modeling for Determining Value and Decision Making*, Chapman & Hall/CRC Press, Boca Raton, FL, 2000.

Marlow, J., *Option Pricing*, John Wiley & Sons, Hoboken, NJ, 2001.

17 Risk Assessment Examples

INTRODUCTION

It is my intention in this chapter to present to the reader some very simple risk-model examples. The purpose of each model is to exemplify an approach that can be used to solve a problem of a specific type. The reader should keep in mind that most real-world risk models are comprised of dozens of variables and, usually, tens if not hundreds of lines of code.

It should also be noted that the examples below do not represent an exhaustive manual on how to solve such problems. The approaches exemplified here were selected for their simplicity and because they are the most generic. Each risk model built presents the builder with a unique set of circumstances that must be handled in a way that is not only efficient, but that is not distasteful to the user.

QUALITATIVE MODEL

The construction of most risk assessment models requires the integration of hard data resulting from real-world measurements or forecasts and soft data that are not expressed quantitatively. An example of a qualitative model may be one in which the answers to questions are statements from which the user can select one or more as the answer to the question.

One approach (one of many, to be sure, but the only one to be outlined here) to treating qualitative data is to rank-order the qualitative answers in order of increasing risk. This is done so that the results of a qualitative assessment may be expressed quantitatively, or so that qualitative data may be integrated with quantitative input. For example, a qualitative risk assessment question and associated answers might be as follows:

Has a security assessment been made in this country?

_____ 1. Yes, very recently — favorable result.
_____ 2. Yes, but some number of years ago — favorable result.
_____ 3. No, but the country has no history of security problems.
_____ 4. No, and the country has a history of security problems.
_____ 5. Yes, and there was indication of security problems.

Because the answers to questions are statements and not numbers, it is generally good practice to convert the statements of increasing risk to a quantitative scale such as 0 to 100. For the question above, there are five answers. Given that we have decided on an output variable scale of 0 to 100 and given that there are 5 answers to this question, the first answer will be assigned a risk value of 0, the second question a value of 25, the third answer a value of 50, the fourth answer a value of 75, and the fifth answer a value of 100. The danger in this methodology is that the 0, 25, 50, 75, and 100 increments do not reflect the real-world relative degrees of risk for the answers. Weights may need to be applied to adjust each answer's risk value.

In the slots in front of each question, users may indicate their confidence in the answer by entering a percentage value. For example, if the user thinks that some time ago a security assessment was performed, but can find no trace of it, he may answer the question thus:

_____ 1. Yes, very recently — favorable result.
_60_ 2. Yes, but some number of years ago — favorable result.
_40_ 3. No, but the country has no history of security problems.
_____ 4. No, and the country has a history of security problems.
_____ 5. Yes, and there was indication of security problems.

This indicates that the user is not sure whether a security assessment was done but that the country does not historically have security problems. On each Monte Carlo iteration, there will be a 60% chance of selecting answer 2 (thus a risk value of 25) and a 40% chance of selecting answer 3 (representing a risk value of 50). On each iteration the risk value from this question is combined with similar information from other questions, such as the following.

What is the quality of the ground transportation infrastructure?

_____ 1. Good to major locations — paved roads.
_____ 2. Roads to major locations — not paved.
_100_ 3. Poor roads everywhere.
_____ 4. No usable roads.

For this question, answer 1 has a risk value of 0, answer 2 a value of 33, answer 3 a value of 66, and answer 4 a value of 100. Again, weights may need to be applied to adjust the risk values. The user has indicated that he is 100% sure that the answer to the question is answer 3. Therefore, on each Monte Carlo iteration, this question will have a risk value of 66.

On a given iteration of this two-question risk model, answer 2 from the first question might be combined with answer 3 from the second question, resulting in a single-iteration overall risk value calculated thus:

$$\text{Total Risk Value} = (25 + 66) / 2 = 45.5 \qquad (17.1)$$

where 2 is the number of questions in the risk model.

Most risk models are composed of a dozen or more questions, each of which can have a unique number of answers. In addition, weights are generally employed, in the form of decimal percentages, to emphasize the relative importance of one

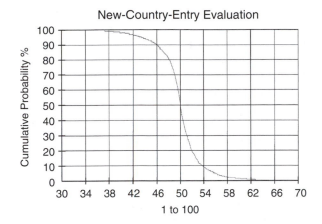

FIGURE 17.1 Cumulative frequency plot of total risk for new-country entry.

question or category of questions to another. For example, we may believe that because we have military or security forces in the area, the security concerns are less important in the country being assessed than are transportation problems. The user may be asked to assign weights (that sum to 1 or to 100) to each question or category of questions. In this case, the user may indicate the following:

- 0.3 = Weight assigned to security problems
- 0.7 = Weight assigned to transportation problems

So, including the weights in the calculation, the total risk is

$$\text{Total Risk Value} = (25 \times 0.3) + (66 \times 0.7) = 53.7 \qquad (17.2)$$

Note that this is a different total risk value than the answer calculated without consideration of weights (45.5). The value of 45.5 is also calculated by using implicit equal weights for both questions (0.5 and 0.5).

Output from a multiple-question qualitative risk model typically appears as shown in Figure 17.1. This risk output curve can stand alone or may be integrated with quantitative information in the same or from other risk models.

One word of warning about such models. This type of model can have serious problems with regard to data scales (see the paper by Pariseau and Oswalt in the selected readings section for this chapter). Readers should be aware of problems concerning the application of ordinal data to risk processes before attempting to build such a model.

SEMI-QUANTITATIVE MODEL

When an ice skater wins a competition, it is because she or he was judged best relative to other participants. So it is for the platform diver, the beauty-pageant participant, and other contestants. In the aforementioned competitions, measurement of a thing is subjective, and such measurements generally are expressed on a scale such as 1 to 10, or 0 to 100, or some other fixed but arbitrary range.

So it is with many aspects of business that may contribute to a risk assessment. For example, we may be interested in evaluating the past performance of a company that we are considering as a partner. Our company may be interested in evaluating our potential partner with regard to many aspects, two of which might be as follows:

- Our relationship with the potential partner's leadership
- The potential partner's environmental compliance record

Our company may decide to judge all aspects on a scale of 1 to 10; 1 indicating good or low risk and 10 indicating poor or high risk. For each question, we may ask the appropriate user (an expert in the subject of the question) for a minimum, most likely, and maximum value on a 1 to 10 scale.

Our expert on the potential partner's leadership indicates that our company has had the best of relationships with the potential partner's present CEO. We have cooperated many times in the past and always to our mutual benefit and satisfaction. However, the current CEO is very near retirement and has warned of it. His successor is known to be much less favorably disposed toward our company and toward such partnerships. The transition from the present CEO to his successor could happen just before or during the formation of the partnership, or it might not happen at all in the time of the partnership. Our expert, therefore, gives us the following values for the risk assessment. Our relationship with the potential partner's leadership:

- Minimum value = 2
- Most likely value = 3
- Maximum value = 8

These values indicate that our expert believes that the most likely situation will be low risk (3), but it could get nasty (8) if the present CEO retires. The minimum value of 2 indicates that our treatment by the potential partner's leadership could be slightly better than expected. We may also ask our expert to supply us with a peakedness value on a scale of 1 to 10 (see Chapter 12 for a discussion of peakedness) to indicate his confidence in his most likely value of 3 (relative to 2 and 8). The expert gives us a peakedness value of 9 indicating high confidence in his assessment. The resulting distribution for this parameter is shown in Figure 17.2.

Our expert on the potential partner's environmental track record has given us the following estimates. The potential partner's environmental compliance record:

- Minimum value = 3
- Most likely value = 6
- Maximum value = 8
- Peakedness value = 5

Our partnership will include some collaboration in foreign countries. Our potential partner has had a relatively good environmental compliance record in the U.S. (a

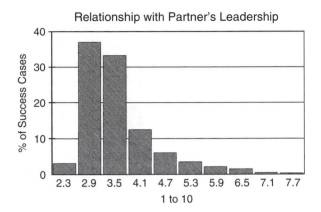

FIGURE 17.2 Frequency plot for evaluation of relationship with potential partner's leadership. Minimum value of 2, most-likely value of 3, maximum value of 8, and peakedness of 9.

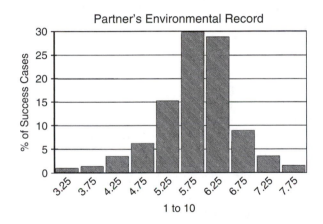

FIGURE 17.3 Frequency plot for evaluation of potential partner's environment record. Minimum value of 3, most-likely value of 6, maximum value of 8, and peakedness of 5.

value of 3). They have, however, run into trouble in several of their non-U.S. projects. In fact, the company was required to terminate one foreign project due to environmental problems (thus, the most likely value of 6 and the maximum value of 8). Because our company would have some control over environmental issues and because only a portion of the partnership venture would be foreign, a peakedness (confidence in the most likely value of 6) of 5 was assigned. The resulting distribution for this parameter appears in Figure 17.3.

These two distributions can be used as input to a risk model that utilizes the very simple equation

$$\text{Total Risk} = (\text{RWPP} + \text{ECR}) / 2 \qquad (17.3)$$

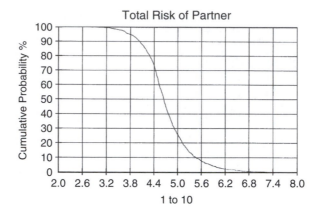

FIGURE 17.4 Cumulative frequency plot of total risk for partnership.

where *RWPP* is the distribution representing our relationship with the potential partner and *ECR* represents the distribution for environmental compliance record. The total risk output distribution appears in Figure 17.4. As with the qualitative model explained in the previous section of this chapter, weights for individual variables or groups of variables could also be applied.

MINI TIME-SERIES MODEL

As outlined in Chapter 14, most projects in a business will have life spans measured in years. Therefore, it is essential that a risk model built to evaluate such an opportunity be capable of assessing not only the present state, but that of years to come as well.

Most real-world time-series models incorporate dozens of variables and, sometimes, hundreds of lines of equations. It is not the point of this section to attempt to impress the reader with intricate models. Rather, it is the aim of the following paragraphs to elucidate just how to approach a time-series risk model. Larger and more complex models adhere to the same principles as are outlined here.

It should be noted that the example presented here is a gross simplification of an actual time-series risk model. In this model, we will consider only two variables: cost and income. Keep in mind that in a more comprehensive model, the variable cost would be composed of other variables such as capital expenses (CAPEX), operating expenses (OPEX), and other costs. Likewise, our variable income would represent the integration of other parameters such as demand, unit price, production rate, and other considerations. Variables such as taxes, discount rate, depreciation, and others typically complicate any real-world analysis.

Our simple model will assess the 5-year project to build a new production facility. Our projected costs, in millions of dollars, for the next 5 years are as follows:

1999	2000	2001	2002	2003
5	3	1.5	0.5	0.5

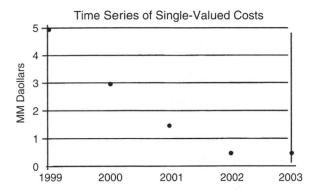

FIGURE 17.5 Time series of single values representing projected yearly costs.

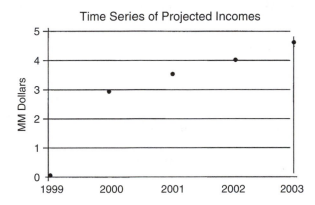

FIGURE 17.6 Time series plot of single values representing projected yearly income.

See Figure 17.5 for a plot of these single values. The $5MM figure for 1999 primarily represents construction costs (in a real-world analysis, construction costs would represent at least one variable). In 2000, we will be operating the plant, but still incurring some capital construction costs. Costs for 2001 are expected to be mainly our operating costs, but the plant is not expected to reach peak operating efficiency until the year 2002. In that year and for subsequent years, our costs are expected to be about $0.5MM per annum.

Our 5-year projection for income, in millions of dollars, looks like this:

1999	2000	2001	2002	2003
0.1	3	3.5	4	4.5

See Figure 17.6 for a plot of these single values. Income in the first year is expected to be zero or near zero, depending upon the date on which we actually begin production.

The value of 3 for the year 2000 reflects the fact that we expect to be producing goods for most of that year. Gauging demand for similar products, market share,

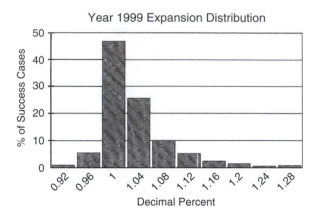

FIGURE 17.7 Year 1999 cost expansion distribution.

competition, regional income levels (all of which would be separate variables in a comprehensive analysis), and so on, we expect our gross sales income to rise by $0.5MM per year.

We have built plants like this one in the past. Our past experience tells us that our first-year costs could be about 10% less than our proposed $5MM value and could exceed the $5MM value by as much as 30% due to labor problems. Therefore, in our risk model, we will, on each iteration (each iteration will treat all 5 years), multiply the $5MM value by a distribution we will call expcost99 (1999 cost expansion distribution). The expansion distribution will have a minimum value of 0.9, a most likely value of 1.0, and a maximum value of 1.3. We are fairly confident in our yearly estimate for 1999, so the peakedness will be set at 10 (see Figure 17.7). See Chapter 14 for details on some problems associated with time-series analysis, iterations, and expansion factors.

For 2000, we believe that the costs could again be lower than the $3MM figure by 10% but may exceed the $3MM number by as much as 25%. Therefore, we will, on each iteration of the Monte Carlo process, adjust the $3MM value by multiplying it by a distribution named expcost2000. This distribution will have a minimum value of 0.9, a most likely value of 1.0, and a maximum value of 1.25. We are less confident that the actual value in 2000 will be $3MM, so we set our peakedness to 8 (see Figure 17.8).

The $1.5MM cost projection for the year 2001 is, relative to 1999 and 2000, a fairly firm number because it is based primarily upon operating costs for plants like this one that we have built previously. Costs are expected to be as much as 5% less than the projected $1.5MM value and might exceed that value by as much as 10%. A distribution named expcost2001 with a minimum of 0.95, a most likely value of 1.0, and a maximum of 1.1 will be used to convert the $1.5MM coefficient to a distribution (see Figure 17.9). A peakedness of 9 will be used.

The $0.5MM figures for the years 2002 and 2003 are expected to vary by no more than plus or minus 5%. A single expansion distribution called expcost023 with

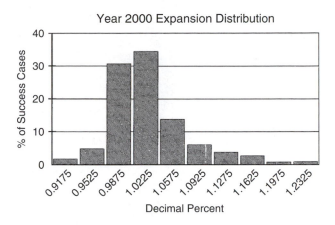

FIGURE 17.8 Year 2000 cost expansion distribution.

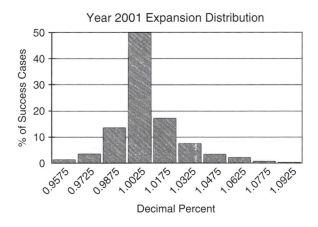

FIGURE 17.9 Year 2001 cost expansion distribution.

a minimum of 0.95, a most likely of 1.0, and a maximum of 1.05 will be used to modify the $0.5M values (see Figure 17.10). Peakedness for this distribution was set to 10.

Our projected gross income for 1999 is expected to be $0 or near $0. To expand our $0.1MM value into a distribution, we will multiply the $0.1MM number on each iteration by a truncated distribution. The minimum and most likely values of the truncated distribution will be 0 and the maximum will be 1.01. We named this distribution expinc99. Peakedness was set to 10 (see Figure 17.11).

For all other years, we expect the income to vary by plus or minus 10%. Therefore, we will create a common expansion distribution called incexp with a peakedness of 10, a minimum value of 0.9, and a maximum value of 1.1 (see Figure 17.12).

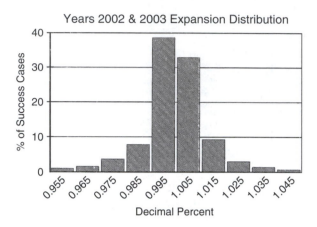

FIGURE 17.10 Cost expansion distributions for the years 2002 and 2003.

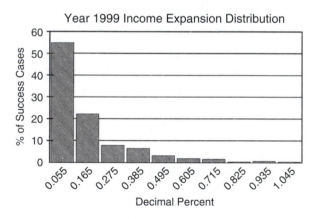

FIGURE 17.11 Year 1999 income expansion distribution.

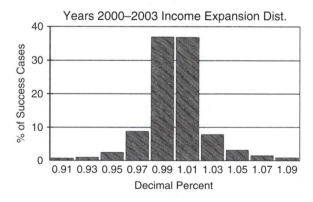

FIGURE 17.12 Income expansion distribution for the years 2000 through 2003.

Total profit for the project would be calculated by the following equations:

$$n = 0;$$

$$costexp\{0\} = expcost99;$$

$$costexp\{1\} = expcost00;$$

$$costexp\{2\} = expcost01;$$

$$costexp\{3\} = expcost023;$$

$$costexp\{4\} = expcost023;$$

while n < PERIODS do

$$total\_cost\{n\} = cost\{n\} * costexp\{n\};$$

if n = 0 then

$$total\_inc\{n\} = income\{n\} * expinc99;$$

else

$$total\_inc\{n\} = income\{n\} * select(expinc,1);$$

endif

$$profit\{n\} = total\_inc\{n\} - total\_cost\{n\};$$

$$n = n + 1;$$

enddo   (17.4)

The variable PERIODS simply holds the number of periods in this analysis which, in this case, is five. The { } syntax simply allows us to reference a specific period (year, in this case). The select function allows us, on each iteration, to sample the expinc distribution one time per period. This allows us to have variability in the expinc variable across periods, not just across iterations.

The final profit plot can be seen in Figure 17.13. As explained in Chapter 12, each of the individual vertical bars can be expanded to display the distribution it represents. If warranted, an NPV (or IRR, or whatever financial measure) function could be called at the end of the program to calculate a net present value for the project.

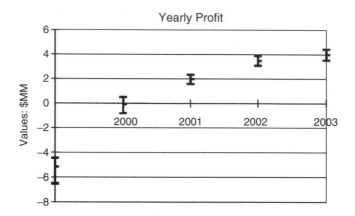

FIGURE 17.13 Time series plot of profit.

FATE-TRANSPORT (PROCESS/HEALTH) MODEL

Fate-transport or process/health models are those that attempt to model movements of substances and related exposure considerations comprehensively. Most fate-transport models attempt to emulate potential human exposure to harmful (or harmful levels of) chemicals or other environmental factors. Typical models and equations attempt to calculate the level of exposure over time. Exposure in humans typically, but not necessarily, falls into several categories. These are as follows:

- Ingestion (including soil, groundwater, food, etc.)
- Inhalation
- Dermal (skin) exposure

The simple example given here is one in which ingestion of a carcinogen will be estimated. The variables in the example will be as follows:

- IR = Water ingestion rate (liters/day)
- CW = Chemical concentration of contaminant in water (decimal percentage)
- EF = Frequency of exposure (days/year)
- ED = Duration of exposure (years)
- AT = Averaging time (days)
- BW = Body weight (kilograms)

From these factors the model will calculate

$$INTAKE = \text{Ingestion of contaminated water (mg / kg - d)} \qquad (17.5)$$

The equation to be used is

$$INTAKE = (CW \times IR \times EF \times ED) / BW \times AT) \qquad (17.6)$$

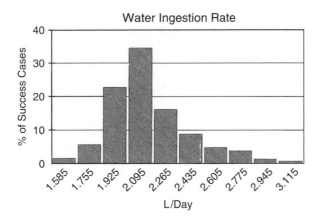

FIGURE 17.14 Distribution of water ingestion rate.

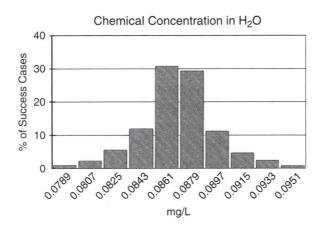

FIGURE 17.15 Concentration of chemical in water.

In the typical deterministic risk assessment, a single value for INTAKE would be calculated from single-valued input parameters. Using the stochastic approach, distributions for IR (Figure 17.14), CW (Figure 17.15), EF (Figure 17.16), ED (Figure 17.17), AT (Figure 17.18), and BW (Figure 17.19) will be used to calculate an output distribution INTAKE (Figure 17.20).

Uncertainties associated with fate-transport-type models are expressed differently than in typical probabilistic models in which uncertainty is likely to be expressed as a percentage or probability and in which the uncertainty actually is a variable in the probabilistic equations. In a process/health-type model, uncertainties generally are

- Listed as uncertainties with sentences describing the source, magnitude, and effect
- Related to gaps in the data

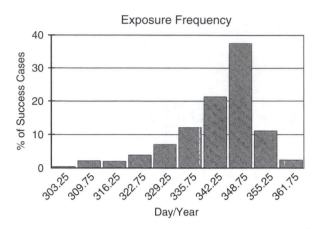

FIGURE 17.16 Distribution of frequency of exposure.

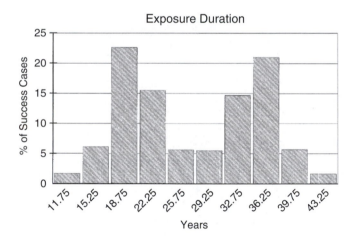

FIGURE 17.17 Distribution for duration of exposure.

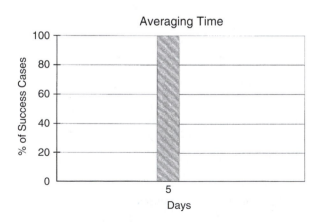

FIGURE 17.18 Single-valued averaging time.

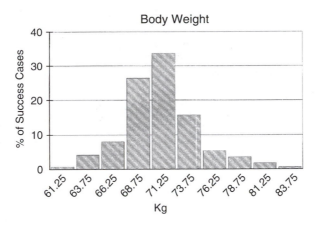

FIGURE 17.19 Distribution of body weight.

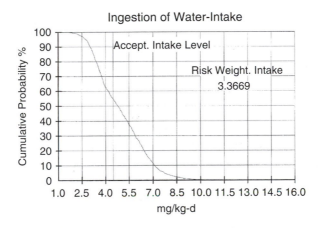

FIGURE 17.20 Cumulative frequency curve of contaminated water intake.

- Related to measurement errors
- Related to sampling errors
- Related to the use of summary statistics or tabular data
- Related to variability in accepted models
- Related to conservation in the models

Fate-transport models are highly complex affairs and do not generally lend themselves to sensitivity analysis as described in this book. Models and equations must be run through several times (changing the coefficient for some parameter each time) to determine the effect of the parameter change on the model result. This is as it should be, given the complexity of the models and the built-in uncertainties. Application of probabilistic techniques to fate-transport-type models can have some beneficial results, but the application is limited due to the nature of most of these models.

DECISION-TREE CONVERSION

Decision trees are popular as a means of generating a deterministic result that incorporates probabilities. Although decision trees do not actually decide anything, they are a very useful decision-making aid.

The visual and graphical nature of a decision tree facilitates discussion of a problem. It forces problem-solving participants to visualize the problem as a more or less linear process from a decision point to one or more consequences. This is a significant step in the process of organizing and diagramming a solution to the dilemma.

In addition, the decision tree allows us to take a step into the probabilistic world. This is not so important a point with regard to calculations or resulting answers. It is a great leap forward, however, for people to begin to think in probabilistic terms and to begin to grapple conceptually with the quantification and documentation of uncertainty.

For all its stalwart features, the decision tree is, in the end, a deterministic vehicle. Some software packages have attempted to imbed the decision tree in a stochastic framework. The meld of these two philosophies often results in either a less-than-potent risk assessment package or one that is unacceptably awkward to use.

My philosophy through the years has been to let decision trees be decision trees and risk packages be risk packages. This has stood me in good stead for several reasons. First, the decision tree structure is an excellent means of directing the evolution of a solution to a problem. Its graphical nature is ergonomic and easy to understand. The decision tree also facilitates determination of where in the problem-solving scheme it is important to incorporate probabilities and which, if any, variables should be represented as distributions. Having thus decided, it is an easy and practical practice to convert the decision tree to a full-blown, if you will, probabilistic model.

The first step in the conversion process is, of course, to generate the decision tree. In Figure 17.21 is depicted a very simple decision tree that is intended to help us decide whether to build a plant. In this admittedly oversimplified example, the go/no-go decision on the plant hinges on whether we will have to pay too much for labor.

Contract negotiations with the labor union are in progress. There is a possibility that the negotiations will go swimmingly and we will realize labor costs well within our fiscal limits. However, things may not go so well. We might experience cost increases due to relatively minor changes to the existing contract, or we could realize significant cost increases from a completely new contract.

If costs are prohibitive, we may decide not to build the plant. This decision is not without cost. Payments to partners, reclamation expenses associated with property-ownership transfer, legal costs, and other penalties could make the do not build decision, at this point, relatively costly.

In our problem-solving sessions, we have described the decision tree seen in Figure 17.21. We have concluded that all variables on the diagram need to be represented by distributions. The distributions representing total costs including those associated with no labor problems, minor labor problems, and a totally rene-

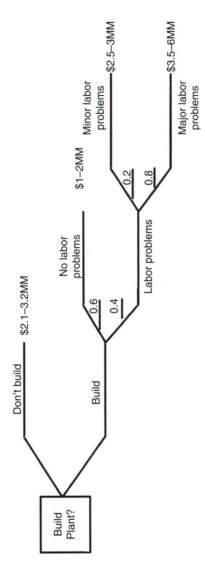

FIGURE 17.21 Decision tree contrasting costs associated with building a plant and possible incurring labor problems with the costs of not building the plant.

FIGURE 17.22 Distribution of costs associated with no-labor-problem scenario.

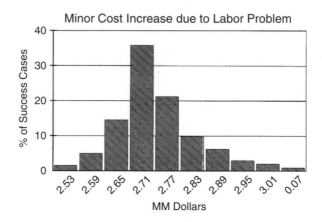

FIGURE 17.23 Distribution of costs associated with minor-labor-problem scenario.

gotiated contact are shown in Figure 17.22, through Figure 17.24, respectively. The projected cost of not building the plant is shown in Figure 17.25.

The simple equation that could be used to convert the decision tree to a Monte Carlo model is

$$\text{buildplant} = (0.6 \text{ x noprob}) + (0.4 \text{ x } ((0.2 \text{ x minorprob})$$

$$+ (0.8 \text{ x majorprob}))) \qquad (17.7)$$

where the variables noprob, minorprob, and majorprob respectively represent the distribution of costs associated with no labor problems, minor labor problems, and totally renegotiated contract. The distribution of costs resulting from this Monte Carlo model is shown in Figure 17.26.

We may be interested in seeing what difference is between building and not building the plant in the Monte Carlo model. If so, we might use the simple equation

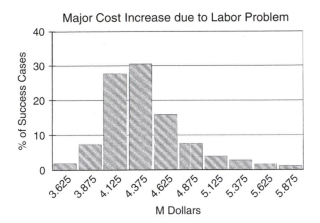

FIGURE 17.24 Distribution of costs associated with a completely new contract.

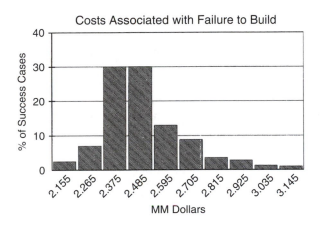

FIGURE 17.25 Distribution of costs associated with failure to build the plant.

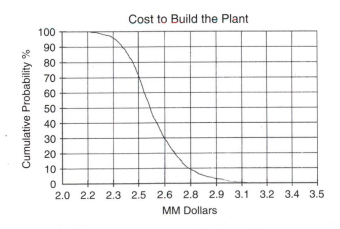

FIGURE 17.26 Cumulative frequency of costs associated with building the plant.

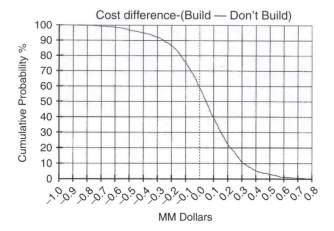

FIGURE 17.27 Cumulative frequency curve of the difference between the costs of building the plant versus not building the plant.

$$\text{diff} = \text{buildplant} - \text{dontbuild} \tag{17.8}$$

This equation will subtract, on each Monte Carlo iteration, the randomly selected cost of not building the plant (the dontbuild variable) from the randomly selected cost of building the plant. The result of the calculation is shown in Figure 17.27.

It should be noted that in this simple example we have used distributions to represent dollar consequences, but have resorted to deterministic estimates of probability. It is a simple matter to represent each probability pair (or group) as dependent distributions. Dependence is necessary because the probabilities need to sum to 1 (or 100, or whatever). So if we were to represent the probability of one branch of the decision tree with a range of 0.2 to 0.6 and on a given iteration the value of 0.35 was selected, then the probability for the complementary branch for that iteration would be 0.65. Note also that this process, in its simplest form, requires that a hierarchy be established for the tree branches. That is, the probability associated with one branch is randomly determined and the other is subject to that randomly determined value. Logic often is added to a model to select randomly, on each Monte Carlo iteration, which branch probability will be drawn from the distribution of probabilities for that branch. The stark difference between utilizing decision-tree probabilities as multipliers vs. as "traffic cops" is demonstrated in a real world example in *Risk Modeling for Determining Value and Decision Making* by this author. See Selected Reading at the end of this chapter.

LEGAL MODEL

Risk assessment models are utilized in legal departments and at law firms to address a wide range of problems. The primary purpose for undertaking a legal risk assessment, of course, is to aid attorneys in making decisions. Equally important, however, are the communication, consistency, and documentation aspects of legal risk assessment.

As a legal decision-making tool, risk modeling is a potent supplement to an attorney's knowledge, intuition, and experience. Typical of the types of questions answered are the following:

- Do we settle this case or pursue litigation?
- If we decide to settle, for how much should we settle?
- If litigation is necessary, what type might be pursued (arbitration, jury trial, adjudication, etc.)?
- What is the probability of a favorable outcome and what will it cost?
- What will be the nonjudgment financial impact (damage to reputation, lost sales or business opportunities, etc.)?
- How do we equitably distribute case loads among staff members?

Although the making of such decisions is a common and integral part of the legal profession, most attorneys find the real power of risk modeling lies in its communication aspects. Prior to the application of risk technology and models, an attorney's conversation with a client would be rife with phrases such as the following:

- We think there is a significant chance of …
- The judgment is likely to be of fairly good size …
- We think there is a better than average chance …

With the advent of risk modeling the use of such nebulous verbiage is on the wane. A parameterized legal risk model provides specific items (variables if you will) that form the basis of a decision. In addition, each of the items requires qualitative, semi-quantitative, or numerical input from both the attorney and the client. Models that serve as the focus of conversations give the client a better understanding and feeling for the decision-making process. Dialogue also gives the attorney a much clearer view of the client's situation, limits, and preferences. At the end of such conversations, and after running the risk model containing the discussed and agreed-upon inputs, the client can be presented with the entire range of possibilities and the associated probabilities.

Prior to the use of legal risk models, an attorney might advise a client thus: "Mr. Smith, we think that this settlement is likely to cost you about $4 million." When settlement is finally reached and costs Mr. Smith $5.2 million, Mr. Smith is displeased, to say the least. Using risk-model output, the presettlement conversation with Mr. Smith is more likely to be as follows:

"Mr. Smith, we put all of the values and judgments that we agreed on into the risk model. Our uncertainty about several of the input parameters was represented as a distribution of values in each case. You can see from the resulting risk assessment plot that there is a small chance that the settlement could cost as little as $3 million. The most likely outcome is that it will cost around $4 million. However, I would like you to notice that the plot shows that there is a small but significant chance that the settlement could cost as much as $12 million. The $12 million figure can be traced back to the input parameters…." Now, when the actual cost comes in at $5.2 million, Mr. Smith is not only relieved (rather than perturbed), but understands the circumstances that culminated in the $5.2 million figure.

Sensitivity analysis is also an integral part of legal risk assessment (see Chapter 15). A sensitivity analysis of a given legal case can indicate to clients which parameters and aspects associated with a litigation are most significant or dire. Such analyses also point out to attorneys which facets of a case deserve the most attention. Much time, effort, and money can be saved by focusing efforts on significant parameters and, equally, by not wasting time tending to items of paltry importance.

Although the aforementioned attributes of legal risk assessment are important, one of the greatest advantages to the use of risk models in the legal profession is associated with consistency. An initial risk model built for a specific legal purpose will integrate the set of parameters outlined in the model-building discussions. A list of parameters generated by this preliminary dialogue is almost always augmented by experience over time. With repeated application of the model, a comprehensive and concise set of parameters results. Such a model should not be seen as stifling flexibility or creativity or as cramping someone's style. Each attorney can contribute to the model those attributes that have personal significance. A well-constructed legal model should allow various users to exercise the required parts of the model in addition to those parts that represent their personal style and flair.

A risk model, although flexible, serves as a consistent checklist of considerations so that no significant points are accidentally overlooked. A model also can offer a consistent means of combining input parameters (flexibility can be built in here, too) and of presentation. Department- or firm-wide consistency of understanding and terminology is of great benefit. Models such as that described here can be invaluable when, for example, an attorney new to the firm or department has to be brought up to speed concerning the way a particular problem is handled.

Yet another paramount benefit of legal risk assessment is that of documentation and archiving. As with most risk models, it often is most valuable to know why a person input the data that they did. A well-constructed risk model should provide for the capture of such information. In addition, the model should offer some means of downloading model input and output data to a database. Such archived information can be invaluable in solving future problems and can be instrumental in forgoing the age-old problems of relearning how to approach a problem and of making the same mistakes over and over again.

The legal risk model presented here is a simple sketch of a generic legal model. Any actual model would, of course, be much more detailed and would address a specific legal problem.

In Figure 17.28 is depicted the contributing factor diagram for this fictitious model. The answer variable in the lower right corner of the diagram indicates that we are interested in determining the difference between the cost of settling a case versus pursuing litigation. Our model is composed of the following simple parameters:

- Probability of a favorable litigation result (expressed as a decimal between 0 and 1). Variable name = LP.
- The internal costs associated with the litigation option. Variable name = LIC.
- The external fees realized from pursuing litigation. Variable name = LEF.
- The settlement amount. Variable name = SA.

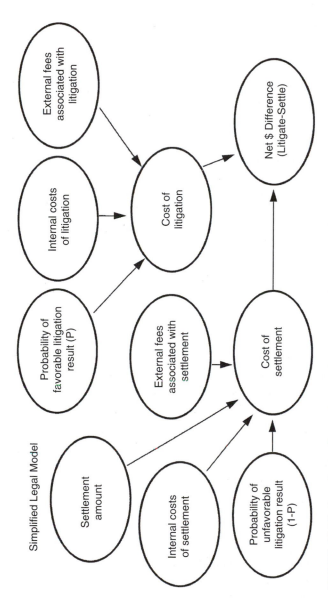

FIGURE 17.28 Contributing factor diagram for a very simple legal model.

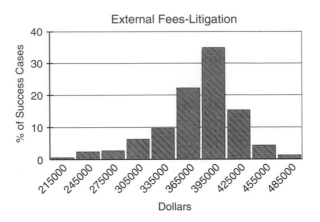

FIGURE 17.29 Frequency plot of external fees associated with litigation.

- The internal costs associated with settlement. Variable name = SIC.
- The external fees realized from settlement. Variable name = SEC.

Our simple set of equations would be

$$TLC = LP \times (LIC + LEF)$$

$$SP = 1 - LP$$

$$TSC = SP \times (SA + SIC + SEC)$$

$$DIFFERENCE = TLC - TSC \qquad (17.9)$$

In the equations, TLC is the total litigation cost, SP is the probability of settlement, TSC is the total settlement cost, and DIFFERENCE is the difference between the cost of litigation and the cost of settlement.

Through conversations with the clients, attorneys have determined the following:

- Litigation external fees will be a minimum of $200,000, most likely $400,000, and a maximum of $500,000.
- Litigation internal costs will be a minimum of $1,000,000, most likely $3,000,000, and a maximum of $4,500,000.
- Settlement external fees will be a minimum of $50,000, most likely $75,000, and a maximum of $100,000.
- Settlement internal costs will be a minimum of $100,000, most likely $200,000, and a maximum of $300,000.
- Settlement amount will be a minimum of $5,000,000, most likely $10,000,000, and a maximum of $13,000,000.
- The probability of a successful litigation is 75%.

Plots of litigation external fees, litigation internal costs, settlement external fees, settlement internal costs, and settlement amount are shown in Figure 17.29 through Figure 17.33. In Figure 17.34 and Figure 17.35, respectively, are plots of the total

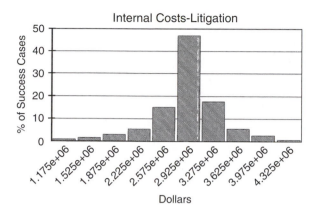

FIGURE 17.30 Frequency plot of internal costs associated with litigation.

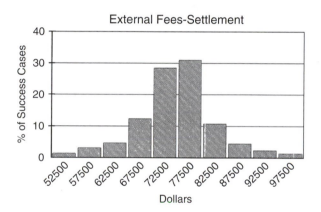

FIGURE 17.31 Frequency plot of external fees associated with settlement.

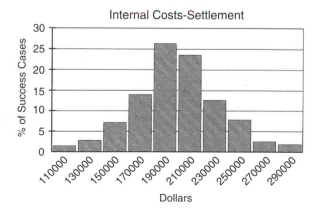

FIGURE 17.32 Frequency plot of internal costs associated with settlement.

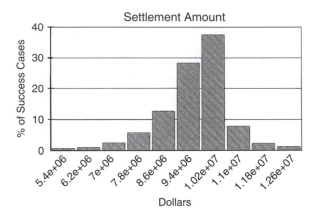

FIGURE 17.33 Frequency plot of settlement amount.

FIGURE 17.34 Cumulative frequency plot of total cost of litigation.

FIGURE 17.35 Cumulative frequency plot of total cost of settlement.

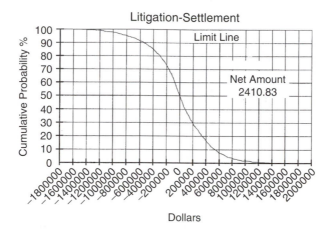

FIGURE 17.36 Cumulative frequency plot of litigation cost minus settlement cost.

cost of litigation and the total cost of settlement. In Figure 17.36 is a plot of the difference between litigation cost and settlement cost.

It can be seen from the information in Figure 17.36 that the probable cost of litigation is nearly the same as the probable cost of settlement. The net amount shown in Figure 17.36 is the net risk-weighted-difference cost of our case. The net value subtracts the probable amount less than 0 (in this case) from the probable amount greater than 0 (see Chapter 15). The net value is $2410.83. This indicates that the settlement option is just slightly better than the litigation option. Given the settlement amounts involved and other values of relatively great magnitude, the $2410.83 value indicates that the two options are essentially the same. To make the decision with regard to settlement and litigation, attorneys might now consider intangible aspects of the case such as the personalities of judges, opposing attorneys, and other elements. They also may opt to add additional considerations to the risk model in an attempt to determine more concisely which option is the better.

A sensitivity plot might also be used to resolve the issue. In Figure 17.37 is shown the sensitivity plot (see Chapter 15) for this case. The sensitivity plot indicates that the most significant variable in our analysis is the internal costs associated with litigation. If attorneys can reduce the uncertainity associated with these costs, the analysis might favor litigation. Note that the influence of the amount of settlement is secondary to the litigation internal costs. This is in spite of the fact that the settlement costs are of greater absolute magnitude.

COMPREHENSIVE RISK ASSESSMENT EXAMPLE

Ben is the general manager in charge of construction and operation of the new chemical plant. Construction of the plant is to begin in less than a year. Ben has called together the staff to begin preliminary discussions. Ben and his staff decide to enlist the help of a risk process facilitator to aid in planning and organizing this project.

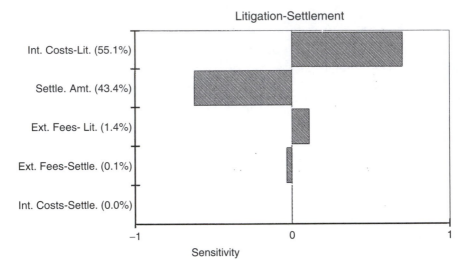

FIGURE 17.37 Sensitivity plot of litigation cost minus settlement cost.

Ben next holds a facilitator-led brainstorming session with his group. The goal of the brainstorming session is to define just what question(s) will be answered by the risk assessment (i.e., just what do they want to know — just what will be calculated by the risk model?). During the brainstorming session, the facilitator leads the discussion and records all pertinent ideas put forth. The simple rules for conducting a brainstorming session are as follows:

- Gather as many participants or stakeholders as is practical
- Record all pertinent ideas
- No criticism

At the conclusion of the brainstorming session, the list of proposed issues is as shown in Table 17.1.

Although the risk model may calculate the answers for several of the items in the brainstorming list, the facilitator realizes that the group should focus on just one question to begin with. The risk model can be expanded to include other items after the risk-model foundation is laid for the most important aspect of the study.

To help in the selection of the primary question to be answered by the risk model, the discussion leader decides to use a hierarchy process. To facilitate this process, the discussion leader distributes to each member of Ben's group two sheets of paper. One paper is the list of brainstorming items in Table 17.1. The second paper is divided into two sections: a not considered now (NCN) section at the top and a FOCUS section at the bottom. Each member of the group is asked to put the number of each statement in Table 17.1 into one of the two sections (NCN or FOCUS items).

Table 17.2 indicates the results of the voting. As can be seen from the table, each statement received at least one vote. This is typical because the person who offers a statement generally votes for it to be a FOCUS item. There are 20 people in the voting group. As a rule of thumb, the facilitator generally considers a statement

TABLE 17.1
Brainstorming Ideas

1. What will be the yearly chemical production from the new plant?
2. What are the plans of our competitor chemical producers?
3. We need to convince upper management that we really need the new chemical plant.
4. What will be the cash flow from the new chemical plant?
5. We need to produce our chemical at a competitive price. How will we do that?
6. What will be the NPV and IRR of the new plant project?
7. This project has to be completed quickly enough so that our competition does not build a similar plant first.
8. What are the environmental constraints on chemical plant production in the new plant area?
9. How will this project compare with other projects competing for corporate funding?
10. We need to build a plant of sufficient efficiency to maintain our market share. How will we do that?
11. What will be the construction costs for the new plant?
12. We should use in-house engineers to design the plant rather than contract engineers.
13. Will we be able to transport the chemical through all states?

TABLE 17.2
Not Considered Now and Focus-Item Votes

Statement #	NCN # Votes	FOCUS # Votes
1	19	1
2	12	8
3	10	10
4	9	11
5	19	1
6	7	13
7	19	1
8	17	3
9	8	12
10	13	7
11	19	1
12	19	1
13	17	3

to be a serious FOCUS item if it receives half or more of the votes. In this case, the statements that received ten or more FOCUS votes are as follows:

- Statement #3 — We need to convince upper management that we really need the new chemical plant.
- Statement #4 — What will be the cash flow from the new chemical plant?
- Statement #6 — What will be the NPV and IRR of the new plant project?
- Statement #9 — How will this project compare with other projects competing for corporate funding?

The facilitator lists these four statements on a flip chart and leads the group through a discussion of the items. In the discussion, the group concludes that statement #4 is covered by statement #6 because NPV and IRR are calculations that consider cash flow. After much discussion and debate, it also is decided that what will convince upper management that the plant is needed (statement #3) is the fact that we can produce and sell economic quantities of our chemical. This will be reflected in NPV and IRR values, and so proponents of statement #3 agree that favorable NPV and IRR values resulting from the risk model will suffice. Similarly, proponents of statement #9, after much discussion, agree that management likely will use NPV and IRR as their main project-ranking tool and, therefore, the calculation of NPV and IRR likely will satisfy their concern about competing corporate projects.

The facilitator realizes that he is lucky this time. In past sessions with other groups, much more contentious discussion ensued that prompted several rounds of voting. In such sessions it has been necessary to have the group vote a second time on only the statements that were selected as FOCUS items in the first voting round (statements 3, 4, 6, and 9 in the current session). In some situations, even after several rounds of voting, more than one statement persisted in the FOCUS group. In such situations, the facilitator has had to have the group decide whether it truly had multiple problems to solve and whether more than one risk model needed to be built to solve them.

In this case, using the hierarchy method, the group settles on the question "What will be the NPV and IRR of the new plant project?" as the first question they will attempt to answer with the risk assessment model. To begin to build the model, the facilitator begins to help the group build a contributing factor diagram. A typical process for building a contributing factor diagram generally posts the answer variable first. Items that lead to the calculation of (items that contribute to) the answer variable are then posted. Figure 17.38 and Figure 17.39 show initial contributing factor diagrams (see Chapter 8 for detailed development of these diagrams).

The group built these diagrams keeping in mind the following:

- The answer variable units are dollars
- All other variables posted will be elements of a system of equations that will eventually calculate dollars

In Figure 17.38 only the answer variable and two major contributing factors are posted. Each of the two contributing factors in Figure 17.38 is discussed by the group and they are expanded into the diagram shown in Figure 17.39.

Although a much more detailed contributing factor diagram could likely be built, the relatively simple contributing factor diagram shown in Figure 17.39 is the one Ben will use for the sake of this example. The simplified contributing factor diagram will be used because it is not practical to present an exhaustive treatment of the problem in this format.

Note that in the contributing factor diagram, units are given for each factor. Note also that some units are, for example, X/Day while others are X/Year. When listing units, always consider what units are most convenient for the user. For example, when asking how many miles a person puts on their car, it is most likely that they

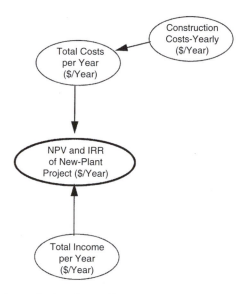

FIGURE 17.38 Initial contributing factor diagram.

will respond with a value like 12,000 per year. However, when asking someone how much they spend for lunch, they are more likely to indicate $6.00 per day.

In the same vein, while some variables are broken down into component parts on the contributing factor diagram, the component parts may not be presented in the risk model. For example, in the contributing factor diagram, Utility Costs is composed of Kilowatt Hours per Year and Cents per Kilowatt Hour. It will be decided in our model that it is easier for the user to estimate the total utility costs per year rather than to guess at kilowatt-hour values.

In addition, it may be decided that we should move some of the factors into a separate model. For example, we may decide to make a separate construction costs risk model, the answers from which would be included in our more comprehensive model.

Ben knows that he has to make preliminary calculations of pertinent costs and incomes. The list of major risk model variables derived from the contributing factor diagram is given below:

- Construction costs
- Hourly wage
- Number of workers
- Utility costs
- Price per ton
- Tons per day
- Discount rate
- Hours per day
- Days per year
- Corporate tax rate

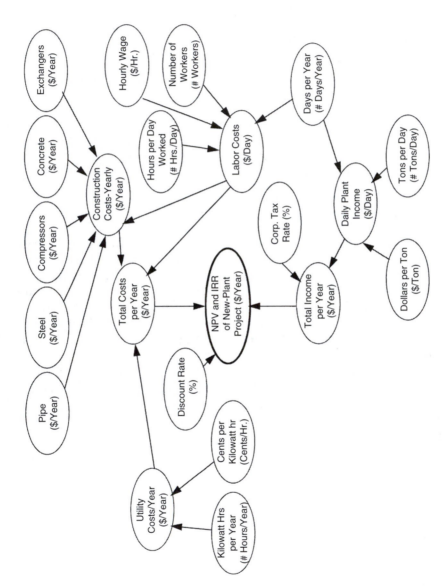

FIGURE 17.39 Completed contributing factor diagram.

TABLE 17.3
Construction Cost Distributions

Component	Minimum (MM$)	Most Likely (MM$)	Maximum (MM$)	Peakedness
Pipe	8.1	9.0	9.9	2
Steel	0.9	1.0	1.1	1
Compressors	10.8	12.0	13.2	4
Concrete	2.25	2.5	2.75	3
Exchangers	2.25	2.5	2.75	2
Labor	22.5	25	27.5	1
TOTALS	46.8	52	57.2	

The team decides that because construction of the plant is to begin soon, the variable construction costs should be set up as a separate model. In initial meetings with the construction cost estimating group, Ben's team and the construction team engage in a ranging exercise for construction cost components. For each component, minimum, most likely, maximum, and peakedness (confidence in the most likely value) values are generated. These are shown in Table 17.3.

The team determines that the peakedness values should be relatively low, indicating their relatively low confidence that the most likely values they had listed would, in the end, actually turn out to be the actual costs. The team was about to enter the totals from Table 17.3 into a spreadsheet/risk model when Jerry, Ben's risk expert, indicated that the summed construction costs should not be entered into the risk model.

Jerry points out that if the individual construction costs are to be entered into the risk model, they certainly could be entered as individual ranges as shown in Table 17.3. However, because they have decided that construction costs are to be summed outside the main risk model, they must be probabilistically summed in a separate risk model. In their risk-modeling system, they build distributions for the construction-cost components shown in Table 17.3 (Figure 17.40 through Figure 17.45) which are summed by the Monte Carlo process using the simple equation

$$\text{total\_const\_cost} = \text{pipe} + \text{steel} + \text{compressors} + \text{concrete} + \text{exchangers} \qquad (17.10)$$

The output distribution for *total_const_cost* is shown in Figure 17.46. This distribution will later be passed as an input distribution to the comprehensive risk model.

As can be seen from the construction cost summing model, the minimum, most likely, and maximum total costs are:

Min.(MM$)	M.L.(MM$)	Max.(MM$)
48.8	52.0	55.4

The minimum and maximum values from the probabilistic sum are significantly different from those resulting from the straight sum. This is so because probabilistic summation of the distributions will not likely combine all of the maximums or all

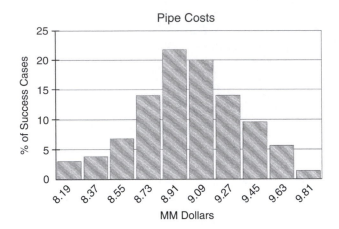

FIGURE 17.40 Pipe-cost distribution used as input for construction-cost model.

FIGURE 17.41 Steel-cost distribution used as input for construction-cost model.

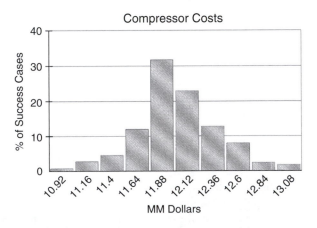

FIGURE 17.42 Compressor-cost distribution used as input for construction-cost model.

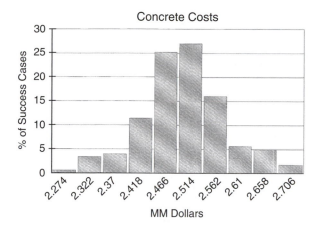

FIGURE 17.43 Concrete-cost distribution used as input for construction-cost model.

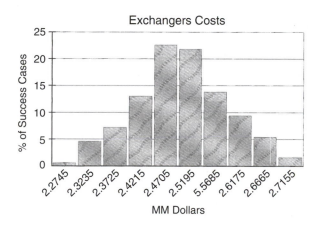

FIGURE 17.44 Exchangers-cost distribution used as input for construction-cost model.

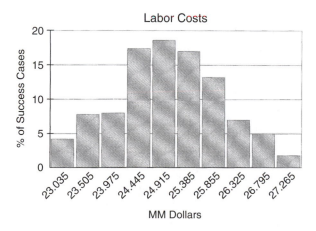

FIGURE 17.45 Labor-cost distribution used as input for construction-cost model.

FIGURE 17.46 Total-construction-cost curve.

of the minimums in a single iteration. This is as it should be and more closely matches the real-world situation.

Ben and Jerry understand that they are in the initial stages of their assessment. This means that they do not have a very good idea regarding the range of costs or the most likely costs. Therefore, Jerry expects that the distributions representing the various parameters will not only be broad, but will have low peakedness values indicating their indecision regarding the most likely values.

In addition, Jerry knows that there are engineers on the evaluation team who are more comfortable using spreadsheets. Jerry is attempting, over time, to wean them away from the spreadsheet risk evaluation mode (a mix of spreadsheets and risk models). He decides, for the time being, to generate all models in spreadsheet/risk model mode. The risk model only and the risk model/spreadsheet/risk model approaches will generate the same results, but will have drastically different run times and user interaction. The approach of using a risk model alone is much more efficient and faster to execute.

In the model, first-year construction costs (FYCC) will be read in as a distribution imported from the constriction-cost risk model. The variables hourly wage (HW), number of workers (NW), utility costs (UC), price per ton (PPT), tons per day (TPD), construction cost multiplier (CCM), and percent of labor force that is construction labor (PLC) will be represented as time-series variables. Discount rate (DR), hours per day (HPD), days per year (DPY), and corporate tax rate (CTR) will be represented as either single values or single-period distributions. The values for these variables will be as shown in Table 17.4. The year-by-year initial values for the time-series variables are shown in Table 17.5. These values have been entered into our spreadsheet and our spreadsheet-driving risk program.

Because the initial stage of the project is primarily a construction phase, the manager whose approval is required to proceed with the project is an engineer. Although known to be a stickler for technical details, she is not known to be so demanding with regard to beginning-stage costs (because we are so early in the

TABLE 17.4
Single-Valued or Single-Period Variables

	Minimum	Most Likely	Maximum
DR (Integer %)		11	
CTR (Decimal %)		0.5	
HPD (Hours)	6.5	7.0	7.2
DPY (Days)	198	200	202

TABLE 17.5
Time-Series Variables

YEAR	1998	1999	2000	2001	2002	2003	2004	2005	2006	2007
HW($)	12	12.5	13	13.5	14	14.5	15	15.5	16	16.5
NW (#)	65	45	30	25	20	20	20	20	20	20
UC (MM$)	0.5	0.75	1	1	1	1	1	1	1	1
PPT (MM$)	0.0002	0.00021	0.00022	0.00023	0.00024	0.00025	0.00026	0.00027	0.00028	0.00029
TPD (Tons)	30	500	550	600	600	600	600	600	600	600
PLC (%)	0.9	0.3	0.05	0.05	0.05	0.05	0.05	0.05	0.05	0.05
CCM (%)		0.3	0.1	0.1	0.004	0.004	0.004	0.004	0.004	0.004

FYCC (MM$) distribution imported from construction cost risk model.

TABLE 17.6
Expansion Distributions for Time-Series Variables

	Minimum	Most Likely	Maximum	Peakedness
HWexp	0.9		1.1	5
NWexp	0.95		1.05	5
UCexp	0.95		1.05	5
PPTexp	0.95		1.05	5
TPDexp	0.9	1.0	1.2	5

planning stage of the project). Most single values in Table 17.5 need to be expressed as distributions. Because we are early in the project and because of the manager's preferences, we will expand each time-series single value for a variable by using a single expansion distribution. The variable-by-variable expansion distributions for the variables found in Table 17.5 are shown in Table 17.6.

The distributions built from the values in Table 17.6 will be used as multipliers for the corresponding single values in each year. This will expand, for example, the 1998 tons-per-day value of 10 from a minimum of 9 to a maximum of 12.

For the sake of simplicity, in this example, we will

- Not apply taxes to negative net income (usually done in an attempt to offset taxes elsewhere in the corporation)
- Not consider depreciation

The spreadsheet used for this model is shown in Figure 17.47. Note that the data in the spreadsheet are not exactly the same as those listed in Table 17.5 and Table 17.6.

In spreadsheets that are driven by risk models, data are overridden on each iteration. Therefore, it is important that the data in the tables be entered only in the risk model. The equations used in the spreadsheet-driving risk model are as follows:

```
c:\directory1\directory2\example.xls;

n = 0;

while n < PERIODS do

CC{n} = FYCC * CCM{n};

HW{n} = HW{n} * select(HWexp,1);

NW{n} = NW{n} * select(NWexp,1);

UC{n} = UC{n} * select(UCexp,1);

PPT{n} = PPT{n} * select(PPTexp,1);

TPD{n} = TPD{n} * select(TPDexp,1);

n = n + 1;

enddo

paramin FYCC B18;

paramin CCM C9:19;

paramin PLC C10:L10;

paramin HW C4:L4;

paramin NW C5:L5;

paramin UC C6:L6;

paramin PPT C7:L7;

paramin TPD C8:L8;
```

    paramin DR B14;

    paramin HPD B15;

    paramin DPY B16;

    paramin CTR B17;

    paramout netincome C19:L19;

    paramout atc C21:L21;

    paramout npv B23;

    paramout irr B24;

    run;

$$irr = (irr * 100) - (dr * 100) \qquad (17.11)$$

where

- c:\directory1\directory2\example.xls; is the path to the Excel spreadsheet
- PERIODS is the number of periods (years) over which we will consider the plant
- {n} is the period (year) index
- select $(x,1)$ is a function that allows resampling of distributions across time periods, not just for each iteration
- paramin and paramout are commands that relate risk-model distributions to spreadsheet cells

After-tax cash flow is shown in Figure 17.48. The resultant distributions for IRR (minus the discount rate) and NPV are shown in Figure 17.49 and Figure 17.50, respectively. Note that the net IRR (minus the discount rate) is just barely positive. The net NPV value is also of small magnitude.

Because the IRR value for the project is positive, the NPV net risk-weighted value is positive (though just barely positive), and the probability of obtaining a positive NPV is around 50, the engineer manager (who has go/no-go authority) allows us to continue with the project. She does this, however, with the warning that the next manager with whom we will have to interact is an economist and, therefore, is likely to have a dim view of a project that has about a 50% chance of a positive NPV.

During the time it has taken Ben and Jerry to complete their initial analysis of the project (about a 6-month period for this project), a lot of things have happened. First, an environmental group has objected to construction of the proposed facility. The environmental group has some misguided ideas about the danger to the community from our proposed plant. The group cannot stop the effort completely, but it can file papers in court to delay the project. The group has, however, agreed to sit down with the company to discuss the realistic expectations with regard to hazards and what the company might do to mitigate these things.

A	B	C	D	E	F	G	H	I	J	K	L
Years		1998	1999	2000	2001	2002	2003	2004	2005	2006	2007
Construction Costs (MM$)		52	15.6	5.2	5.2	0	0	0	0	0	0
Hourly Wage ($/hr)		12	12.5	13	13.5	14	14.5	15	15.5	16	16.5
Number of Workers (Integer)		65	45	30	25	20	20	20	20	20	20
Utility Costs (MM$)		0.5	0.75	1	1	1	1	1	1	1	1
Price per Ton (MM$)		0.0002	0.00021	0.00022	0.00023	0.00024	0.00025	0.00026	0.00027	0.00028	0.00029
Tons per Day (Tons)		10	500	550	600	600	600	600	600	600	600
Construction Cost Mult. (%)		1	0.3	0.1	0.1	0	0	0	0	0	0
Percent Labor Const. (%)		0.9	0.3	0.05	0.05	0.05	0.05	0.05	0.05	0.05	0.05
Revenue (MM$)		0.4	21	24.2	27.6	28.8	30	31.2	32.4	33.6	34.8
Expense (MM$)		0.61232	1.067	1.53352	1.4617	1.38304	1.39672	1.4104	1.42408	1.43776	1.45144
Expenditures (MM$)		53.01088	15.843	5.22808	5.2243	0.02016	0.02088	0.0216	0.02232	0.02304	0.02376
Discount Rate (Decimal %)	0.11										
# Hours per Day (Floating)	7.2										
# Days per Year (Integer)	200										
Corp. Tax Rate (Decimal %)	0.5										
First-Year Const. Cost (MM$)	52										
Net Income Before Tax (MM$)		−0.21232	19.933	22.66648	26.1383	27.41696	28.60328	29.7896	30.97592	32.16224	33.34856
After-Tax Cash ($)		−53.2232	−5.8765	6.10516	7.84485	13.68832	14.28076	14.8732	15.46564	16.05808	16.65052
NPV	−1.311904										
IRR	10%										

FIGURE 17.47 Initial spreadsheet to be driven by the risk model.

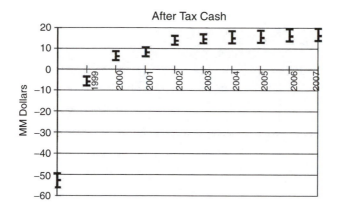

FIGURE 17.48 Time-series plot of after-tax cash from initial risk model.

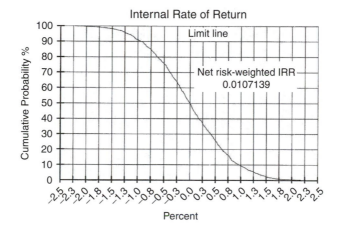

FIGURE 17.49 Plot of internal-rate-of-return distribution from initial risk model.

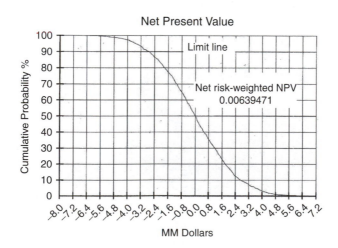

FIGURE 17.50 Plot of net present value from initial risk model.

TABLE 17.7
Distributions for New Parameters

	Minimum	Most Likely	Maximum	Peakedness
PD (decimal %)	0.1	0.15	0.3	10
DT (Days)	30	60	90	5
DPDD ($)	10,000	15,000	20,000	9

	1998	1999	2000	2001	2002	2003	2004	2005	2006	2007
ACC (MM$)	2	1	0.5	0	0	0	0	0	0	0

TABLE 17.8
Time-Series for Increased Production from New Technology

	1998	1999	2000	2001	2002	2003	2004	2005	2006	2007
NTIP (T/D)	0	50	75	85	85	85	85	85	85	85

The new parameters that this development will add to the risk model(s) are probability of delay (PD), delay time (DT), dollars per day delayed (DPDD), and additional construction costs (ACC) associated with mitigation of perceived and real environmental problems (see Table 17.7).

In addition, during the 6-month period, the political problem of zoning has reared its ugly head. The proposed plant site is not too distant from a residential area (thus, the environmental objections). Completely unrelated to the project, the local housing developers are pushing the legislative bodies to include the potential plant site in the residential zone. If the plant site is zoned as residential, the project will be killed. The company is not without influence with the local government. Many jobs will be created if the new plant is built. Millions of dollars per year will be pumped into the local economy by the new facility. Therefore, the company believes that there is only about a 15% chance that the project will be killed altogether. The variable Ben will use to represent the probability of failure due to zoning is PZF.

Corporate intelligence indicates that the competition may be planning to build a plant similar to Ben's to capture the market share that Ben needs for this to be an economic venture. It has been decided that if the competition begins to build a plant prior to our major construction-spending phase (i.e., while Ben is still in the planning phase), then he will not build the plant. The best estimate from the corporation regarding the competition's plant is that there is about a 10% chance that they will beat Ben to the punch. Therefore, there is a 10% chance of abject project failure due to the competition. Ben will use the variable FDTC to represent failure due to competition.

Also during the 6-month period, new technology has been developed that can significantly boost production. The cost of the new technology supplanting older technology will not be significant. (Construction costs for the new equipment and construction costs for the old equipment are about the same.) The projected new technology increased production (NTIP) figures are shown in Table 17.8.

As mentioned previously, the new manager is an economist. This person thinks that he has a better grip on how prices will change from year to year. The manager and a project economist have suggested the expansion distributions for prices shown in Table 17.9.

The new expansion distributions are shown in Figure 17.51 through Figure 17.54. Note the tighter range in the distributions and the generally higher peakedness values (except for the 2001–2007 time span). The restricted range and generally higher peakedness values reflect the economists' confidence in their ability to predict prices in the coming years.

TABLE 17.9
Expansion Distributions for Prices

	Minimum	Most Likely	Maximum	Peakedness
PPTexp 1998	0.9	1.0	1.1	9
PPTexp 1999	0.85	1.0	1.3	7
PPTexp 2000	0.75	1.0	1.4	5
PPTexp 2001–2007	0.6	1.0	1.4	0

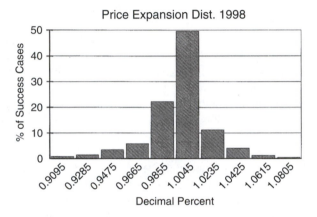

FIGURE 17.51 Expansion distribution for 1998 prices. The deterministic time-series value is multiplied by this distribution to expand the time-series value into a distribution.

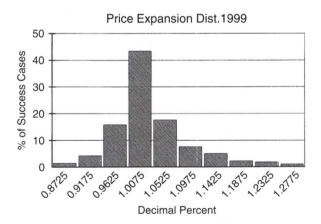

FIGURE 17.52 Expansion distribution for 1999 prices. The deterministic time-series value is multiplied by this distribution to expand the time-series value into a distribution.

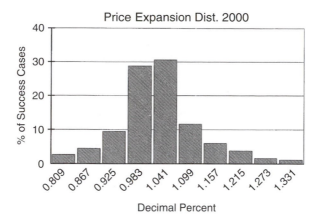

FIGURE 17.53 Expansion distribution for 2000 prices. The deterministic time-series value is multiplied by this distribution to expand the time-series value into a distribution.

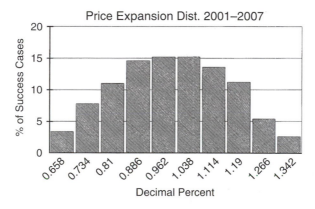

FIGURE 17.54 Expansion distribution for 2001–2007 prices. The deterministic time-series value is multiplied by this distribution to expand the time-series value into a distribution.

The risk model must be modified to reflect the following:

- Probability of environmental delay
- Delay time
- Dollars per day of delay
- Additional construction costs to mitigate environmental problems
- Probability of abject project failure due to zoning
- Probability of abject project failure due to the competition
- New-technology-induced increased production
- New price-expansion distributions

The new set of equations used in the risk model are as follows:

```
c:\directory1\directory2\example2.xls;

n = 0;

while n < PERIODS do

if n = 0 then

        PPT{n} = PPT{n} * PPTexp98;

elseif n = 1 then

        PPT{n} = PPT{n} * PPTexp99;

elseif n = 2 then

        PPT{n} = PPT{n} * PPTexp00;

else

        PPT{n} = PPT{n} * select(PPTexp0107,1);

endif

CC{n} = FYCC * CCM{n};

HW{n} = HW{n} * select(HWexp,1);

NW{n} = NW{n} * select(NWexp,1);

UC{n} = UC{n} * select(UCexp,1);

TPD{n} = TPD{n} * select(TPDexp,1);

NTIP{n} = NTIP{n} * select(NTIPexp,1);

n = n + 1;

enddo

paramin FYCC B18;

paramin CCM C9:19;

paramin PLC C10:L10;

paramin HW C4:L4;

paramin NW C5:L5;

paramin UC C6:L6;
```

paramin PPT C7:L7;

paramin TPD C8:L8;

paramin DR B14;

paramin HPD B15;

paramin DPY B16;

paramin CTR B17;

paramin PD B20;

paramin DT B21;

paramin DPDD B22;

paramin ACC C23:L23;

paramin NTIP C24:L24;

paramout netincome C27:L27;

paramout atc C29:L29;

paramout npv B31;

paramout irr B32;

run;irr = (irr * 100) — (dr * 100); (17.12)

where

- c:\directory1\directory2\example2.xls; is the path to the Excel spreadsheet
- PERIODS is the number of periods (years) over which we will consider the plant
- {n} is the period (year) index
- Select $(x,1)$ is a function that allows resampling of distributions across time periods, not just for each iteration
- Paramin and paramout are commands that relate risk-model distributions to spreadsheet cells

The spreadsheet being driven by the risk model appears in Figure 17.55. The output NPV and IRR plots are shown in Figure 17.56 and Figure 17.57, respectively.

In Figure 17.56, it can be seen that the NPV curve still extends slightly into the negative-dollar range. The extent of the NPV curve into the negative range is much reduced relative to the NPV curve produced by the first model. Although prices have not changed significantly, the increase in production due to the introduction of new technology has had a relatively great positive impact on NPV. The net risk-weighted NPV value reflects this. Note also the Y-axis intercept for the NPV cumulative

A	B	C	D	E	F	G	H	I	J	K	L
Years		1998	1999	2000	2001	2002	2003	2004	2005	2006	2007
Construction Costs (MM$)		52	15.6	5.2	5.2	0	0	0	0	0	0
Hourly Wage ($/hr)		12	12.5	13	13.5	14	14.5	15	15.5	16	16.5
Number of Workers (Integer)		65	45	30	25	20	20	20	20	20	20
Utility Costs (MM$)		0.5	0.75	1	1	1	1	1	1	1	1
Price per Ton (MM$)		0.0002	0.00021	0.00022	0.00023	0.00024	0.00025	0.00026	0.00027	0.00028	0.00029
Tons per Day (Tons)		10	500	550	600	600	600	600	600	600	600
Construction Cost Mult. (%)		1	0.3	0.1	0.1	0	0	0	0	0	0
Percent Labor Const. (%)		0.9	0.3	0.05	0.05	0.05	0.05	0.05	0.05	0.05	0.05
Revenue (MM$)		0.4	23.1	27.5	31.74	33.12	34.5	35.88	37.26	38.64	40.02
Expense (MM$)		0.61232	1.067	1.53352	1.4617	1.38304	1.39672	1.4104	1.42408	1.43776	1.45144
Expenditures (MM$)		55.14588	16.978	5.86308	5.3593	0.15516	0.15588	0.1566	0.15732	0.15804	0.15876
Discount Rate (Decimal %)	0.11										
#Hours per Day (Floating)	7.2										
#Days per Year (Integer)	200										
Corp. Tax Rate (Decimal %)	0.5										
First-Year Const. Cost (MM$)	52										
Prob. of Delay (Decimal %)	0.15										
Delay Time-Eve. Prob. (Days)	60										
$/Day of Delay ($)	15000										
Additional Const. Costs (MM$)		2	1	0.5	0	0	0	0	0	0	0
New Tech. Inc. Prod. (Tons/Day)		0	50	75	90	90	90	90	90	90	90
Net Income Before Tax (MM$)		-0.21232	22.033	25.96648	30.2783	31.73696	33.10328	34.4696	35.83592	37.20224	38.56856
After-Tax Cash ($)		-55.3582	-5.9615	7.12016	9.77985	15.71332	16.39576	17.0782	17.76064	18.44308	19.12552
NPV	4.9069229										
IRR	13%										

FIGURE 17.55 Final spreadsheet to be driven by the risk assessment model.

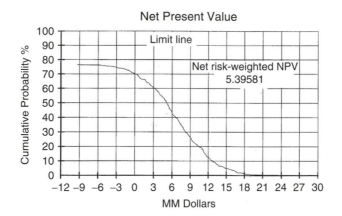

FIGURE 17.56 Final net-present-value distribution.

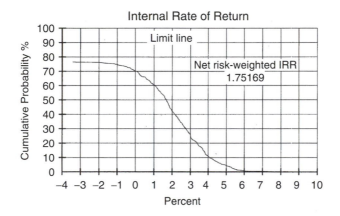

FIGURE 17.57 Final internal-rate-of-return distribution.

frequency curve. The chance of failure from zoning is 15% (0.15). The chance of failure due to the competition building first is 10% (0.1). The total chance of success is given as follows:

$$TCS = (1 - 0.15) \times (1 - 0.1) = 0.85 \times 0.9 = 0.765 \qquad (17.13)$$

This is, of course, a 76.5% chance of success. The net risk-weighted value considers the 23.5% chance of abject failure in its calculation. This same impact can be seen in the IRR cumulative frequency curve. The significantly positive NPV and IRR net risk-weighted values (in spite of the 23.5% chance of abject failure) indicate that the project likely should be pursued.

EVS/EVP EXAMPLE

In the previous example (Comprehensive Risk Assessment Example), I focused on the details of costs, how to represent parameters as ranges, chances of failure, integration of the many parameters, the computer code used to perform the analysis, and many other aspects. In this example, I will forgo much of the detail to concentrate on the results of the assessments and what they mean.

In keeping with the chemical plant theme, the example given here will consider production from such a plant, but this time from two different perspectives. The first assessment example delineated below considers success or failure of a plant from the corporate perspective. The second example demonstrates how a business-unit leader might view the same project. Again, because the previous section delved into much detail, this section will not repeat the dive into minutiae but will take the 30,000 foot view of the problem.

THE CORPORATE VIEW

So, consider that a corporation has multiple investment opportunities, one of which is to build a chemical plant. It will base its decision about which opportunities to fund partly on whether or not the projects are forecast to have positive NPVs (net present values). This, of course, would not be the only criterion upon which a decision would be made, but it is a critical element. Also, just because a project is purported to exhibit a positive economic metric, this does not mean that the project will be carried through to completion — it only means that the project will be allowed to go on to the next stage of evaluation and development. At any subsequent stage and decision point, the project will be re-evaluated and could be cancelled.

From the 30,000 foot (high-level) view, the five essential steps in the evaluation are as follows:

1. Agree on what constitutes a go case (the project can proceed to the next evaluation stage). In this case, we agree that a forecasted positive NPV will be the trigger.
2. Identify all important parameters and discuss and assign uncertainty to all pertinent components.
3. Agree on go/no-go parameters (chances of abject failure).
4. Apply uncertainties and generate fully uncertainty-impacted go NPV value (EVS).
5. If failure modes are pertinent, apply all and generate fully risked NPV value (EVP). If there are no chances of abject failure, then EVS = EVP.

There are many project parameters about which we are uncertain. Just a few are listed in Table 17.10 with the associated ranges. All parameters about which there is uncertainty would be represented in the model by distributions. Only one such distribution — fixed costs — is shown in Figure 17.58 as an example. All such distributions are integrated in our assessment model to produce a probabilistic cash-flow plot and the resulting NPV distribution. Plots of both cash flow and NPV are shown in Figure 17.59 and Figure 17.60.

TABLE 17.10
Just a Few Parameters about Which We Are Uncertain

	Minimum	Most Likely	Maximum
Plant capacity (Tons/Year)	180,000	200,000	220,000
CAPEX Year 1 (MM$/Year)	48	50	52
CAPEX Year 2 (MM$/Year)	45	48.9	50
Fixed Costs (MM$/Year)	11.8	12.5	12.9

CAPEX is capital expenditure and MM$/Year is millions of dollars per year.

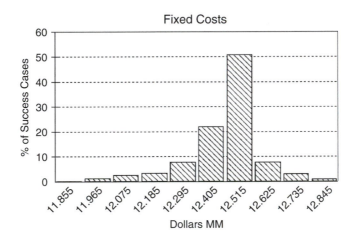

FIGURE 17.58 Distribution of fixed costs for chemical plant.

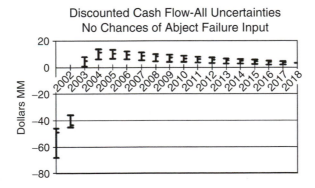

FIGURE 17.59 Plot of chemical plant probabilistic cash flow.

Note that the NPV plot indicates that there is about a 10% chance that the project could have a negative NPV. Therefore, if decision makers decide to go forward, they must do so knowing that, even though the EVS (mean) is positive, there is a chance

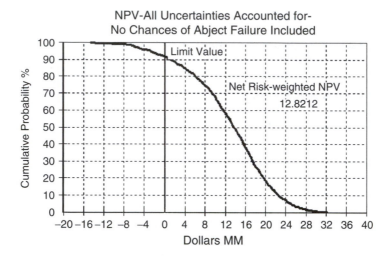

FIGURE 17.60 NPV cumulative frequency plot and EVS value.

that this project could exhibit an NPV that would not pass the go/no-go test. As seen on the plot, the EVS NPV is about $12.8MM. According to our agreed-upon criterion, this is a go situation — the project will likely proceed to the next evaluation phase. Remember, the "S" in EVS indicates that we have been successful in executing the project and *not* that we have executed a successful project.

Now, if there exist any no-go considerations (i.e., chances of abject failure), we should apply them to generate our EVP. Of course, there could be a host of failure situations associated with such a project, but we will here consider only two:

- Wait time. The project might fail if we have to wait, while experiencing significant spending, more than 1 year to begin the project. Delay in issuance of permits and other political, environmental, technical, and legal problems could contribute to delay. We know that the competition is considering building a similar plant to capture the market. If we have to wait more than 1 year to start the project, it is envisioned that the competing plant will be beat us to the market and it will make no economic sense to continue with our project.
- War. Given the political situation in the country in which the plant will be built, we believe that there is some chance that a war will break out sometime prior to or during the plant-construction phase. If war does break out in this time period, the project will be abandoned.

Project analysts have collected data on wait time for all similar projects (by any company) in the part of the world being considered. The analysts have combined this data with the expert judgment of corporate personnel. The plot in Figure 17.61 shows the range of wait times that can be expected and that there exists about a 12% chance that we will have to wait more than 1 year. Similarly, the company has gathered opinions from experts regarding the possibility of war. The consensus was

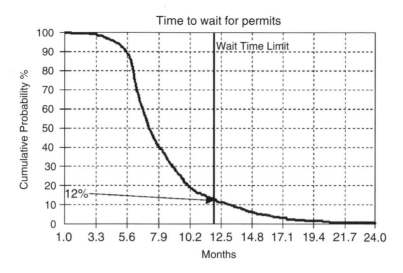

FIGURE 17.61 Wait-time cumulative frequency curve.

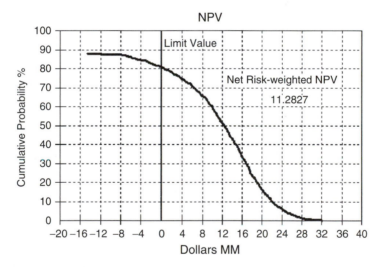

FIGURE 17.62 NPV cumulative frequency curve showing impact of unacceptable-wait-time chance of failure.

that there is about a 20% chance that war will break out sometime in the timeframe being considered.

For the sake of this example, each chance of failure — wait time and war — will have its impact applied separately. In an actual model, unless there was some reason to consider these consequences separately, a model might initially consider the combined impacts of all chances of failure to produce an EVP. Figure 17.62 shows the effect of the 12% chance of an unacceptable wait time. In Figure 17.63 is depicted the combined impact of unacceptable wait time and war.

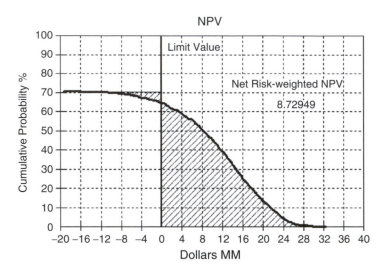

FIGURE 17.63 NPV cumulative frequency curve showing combined impact of unacceptable wait time and war.

Note that the EVP/NPV considering only the unacceptable wait time is about $11.3MM. Taking into account both unacceptable wait time and war drives the EVP/NPV down to about $8.7MM. This is still a positive value, so corporate executives might be inclined to allow the project to proceed to the next evaluation stage.

The final EVP is the value that should be used by corporate executives to compare with EVPs from other projects to help decide which projects are the best investment opportunities. The EVS represents the prize that is possible if all chances of abject failure can be successfully and completely mitigated. To summarize, the EVS/NPV for the project is about $12.8MM. The EVP/NPV considering only the unacceptable wait time is about $11.3MM. The EVP/NPV reflecting both the unacceptable wait time and the possibility of war is about $8.7MM.

THE BUSINESS-UNIT-LEADER VIEW

A business-unit leader (BUL) might be expected to deliver a predetermined volume of product (chemical) in the first year of plant operation. According to the plots shown in Figure 17.64 and Figure 17.65, the first-year yield should average about 135,000 tons. This is the EVS value. It is the value for which we make logistical plans (build pipelines, rent train cars, etc.). It is the value that we supply to the engineers for all aspects of construction because if we are lucky enough to experience a successful plant start-up and successful production, this is the volume we need to be prepared to handle.

To the corporation, failure is defined to be failure to start plant construction or failure to begin to operate a partially completed plant. However, failure to the BUL can be extended to mean failure to deliver expected volumes in the first year of operation of the plant. So, in addition to the plant-stopping chances of failure such

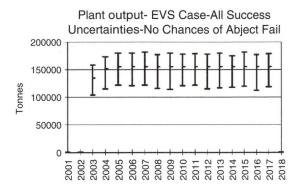

FIGURE 17.64 Time series plot of probabilistic plant production.

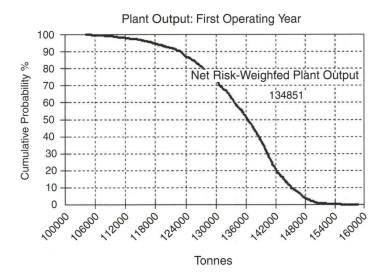

FIGURE 17.65 Cumulative frequency plot of first year production.

as unacceptable wait time and war, the BUL might consider problems that could cause him to fail to deliver the expected chemical production in the first year of operation.

One chance of abject failure for the BUL might be delay in initial production due to working out bugs in new and untested technology incorporated in this plant. Such a delay would not cause the plant to fail, but would certainly cause the BUL to fail to deliver in the timeframe being considered (the first year of operation).

It is estimated that the time required to work bugs out of new technologies might be as little as a few weeks and is most likely to take about 6 months. However, it can be envisioned that the delay could be more than 1 year. It is estimated that there is about a 10% chance that delays will be of sufficient magnitude to cause the BUL to miss his production target in the first year of operation.

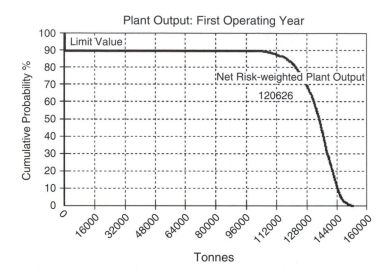

FIGURE 17.66 Production EVP considering only delays due to working bugs out of new technology.

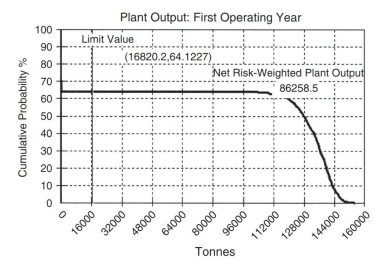

FIGURE 17.67 Production EVP considering all chances of abject failure.

The plot shown in Figure 17.66 indicates that the EVP production considering *only* the impact of delay due to working bugs out of new technologies is about 120,700 tons. This is down from the EVS value of about 135,000 tons. The EVP production considering all chances of abject failure (unacceptable wait times, war, and production delays due to technology bugs) is about 86,300 tons. This value results from the plot shown in Figure 17.67. This value is significantly different from the EVS value of about 135,000 tons.

So, the BUL should use the EVS value to inform engineers, logistical experts, and others regarding how big to build facilities, what volumes should be considered in contracts, and the like. The promise to the corporation for consideration of roll-up in total corporate production should be the final EVP value.

SELECTED READINGS

Finkler, S. A., *Finance and Accounting for Nonfinancial Managers*, Prentice Hall, Englewood Cliffs, NJ, 1992.

Koller, G. R., *Risk Modeling for Determining Value and Decision Making*, Chapman & Hall/CRC Press, Boca Raton, FL, 2000.

Pariseau, R., and Oswalt, I., Using data types and scales for analysis and decision making, *Acquisition Review Quarterly*, 145–159, Spring, 1994.

Rodricks, J. V., *Calculated Risks — Understanding the Toxicology and Human Health Risks of Chemicals in Our Environment*, Cambridge University Press, Cambridge, U.K., 1994.

Index

A

Absolute minimum and maximum values, 169
Access to executives, 48–49
Accuracy and precision, 72–74
Adding variables, 112–114, 122
Analytical decision tree, 144–146
Archiving, 280
Area under cumulative frequency curve, 233–236
Artificial intelligence, 149–150
Asking the right question, 113, 114–115, 253–256, 286–287
Auditing, 61–62
Awareness, 17–18, 32

B

Bar charts, 110–112, 162–163, 173–175
Bayes, Thomas, 139
Bayesian analysis, 137, 138–141
Benchmark, 77
Beta distribution, 185, 189
Big picture view, 32–33, 48
Bimodal distributions, 112, 173, 177, 184–185
Bin size, 173, 175
Black-Scholes model, 245–246
Board of review, 28
Branching, probabilistic, 20, 241–242, 278
Brokerage house, 24, 71
Buckets, 159–161, 173
Budget estimation, 33, 248–252
    risk-weighted scheme, 60–61
Business case for risk process implementation, 42–43

C

Capital project risk model, 26
Cash flow, 65
    comprehensive risk assessment example, 287, 288, 297
    consensus model, 101
    contributing-factor diagram, 119, 121
    financial measures, 103–104, 213, *See also specific measures*

time-series analysis, 213, 214, 215, 218
Categories, 24, 25, 30
    contributing factor diagram construction, 113–114
    discrete distributions, 183
    fate/transport model, 270
    overarching, 103–105
    qualitative model, 261
    specific distributions, 189
Chance of failure, 197–211, 213
    Bayesian analysis, 140
    budget estimation, 250–251
    chain link analogy, 197, 207
    comprehensive risk assessment example, 301, 307
    cumulative frequency curves, 104, 167, 201–203
    defining, 197, 198, 207
    education, 126, 127, 128
    effect on input distribution, 200–201, 203–205
    EVS/EVP concept and, 9, 197, 208–210, 310–311, *See also* Expected value for the portfolio
        business-unit-leader view, 312–314
        corporate view of, 310–311
    facilitator tasks, 28
    multiple output variables, 207
    new technology and, 313
    package shipment rate example, 203–205
    percentiles and, 203
    plant startup example, 207–210
    risk-weighted values, 229, 237
    substituting zeroes work-around, 205–206, 210
    water project example, 198–201
Checklist, 23
Check the box mentality, 63–64, 86
Chi-squared distributions, 189–190
Circular dependence, 224
Clipped distributions, 199, 205
Common factors, 68
Communication, 32, 83–95
    big picture view, 48
    common language, 14–16, 81
    with customers and the public, 19–20
    distribution selection and presentation, 190–194